A Biodynamic Book of Moons

By Dennis Klocek

Published in 1983 by Bio-Dynamic Literature
Wyoming, Rhode Island

Republished 2023
By Soil, Soul and Spirit and Dennis Klocek
Carmichael, California

With support from the Coros Institute at coros.org

Paperback ISBN-13: 979-8-9883689-0-8
eBook ISBN-13: 979-8-9883689-1-5

EARTH WATER AIR FIRE EARTH WATER AIR FIRE

ACKNOWLEDGMENT

I would like to express my appreciation to those who have made this book possible:

To Heinz Grotzke, for his expert guidance, to Anne Marshall for editorial work, to Horace Cartter, for book lay-out and production, and especially to my wife, Barbara, for her unflagging loyalty, support and encouragement.

Dennis Klocek

Contents

EARTH WATER AIR FIRE EARTH WATER AIR FIRE

Preface to the 40th Anniversary Edition

The saying goes that an author's first book contains the seeds of all of the later works. That has proved true in my work on the Book of Moons. The impulse to this book arose out of my admiration for the work of three men. The first, my maternal grandfather, an avid gardener who fertilized his garden with manure he brought home from the mules that pulled the coal cars in the mines where he worked. As a child I wondered what he was looking at when I would see him in his garden standing still and staring at the plants. I often think of him when I find myself doing that same thing in my own garden.

Growing up in New Jersey just across the Delaware river from Philadelphia the mood of colonial America was a strong theme in my life. This was especially so in the work of Benjamin Franklin. His scientific curiosity and his practical work at the printing press was for me developed to a perfect synthesis in his Poor Richard's Almanac. The first attempt I made of a little chap book in that style was the original form of the Book of Moons that I sent as a xerox copy of to Heinz Grotzke who was the editor of the BD press. The little 9x12 hand sewn booklet with potato print margins, pen and ink drawings and hand lettered text must have given him a little laugh when he opened the envelope. He graciously wrote back that if I expanded it, he would be willing to take another look.

A few years later in my studies towards a master's degree in fine arts I came across the color theory of Goethe. His blend of scientific study with the eye of a poet planted the seed of an expanded version of the chap book with woodblock prints and poems in the style of Franklin's rhyming wisdom. A turning point for me was the chance discovery of the carcass of a large white bird in a bush I found on a morning walk. The head was gone as was the breast meat but the wings were intact and full of large handsome quills. I took them and made them into pens and began to create poems for the expanded version out of the ideas for each month. In the same woods a large choke cherry tree had fallen and dried to a perfect seasoning without touching the ground and rotting. With an inherited two-man saw the wood was converted into planks that became the woodcut images for the text. Several years later the book was put to press by Mr. Grotzke.

The themes of the Book of Moons have stayed with me my whole life. They are the interest in star wisdom as a lens for nature study and weather, the mysteries of plant growth in the context of the passing of the seasons, and the importance of an artistic approach to the study of nature as the basis for a new healing approach to the challenges now facing the human soul.

Dennis Klocek
January, 2023 Carmichael, California

EARTH WATER AIR FIRE EARTH WATER AIR FIRE

Introduction

The term *Earth Logos* comes from the Greeks and means *Earth Being*. To the Greek mind this not only referred to the Being of the earth but to the knowledge received from the Earth Being. This knowledge is freely given to anyone who touches the soil. At the time that the earth is touched, understanding is given on how to care for the plants and animals found on and in the earth's body. By eating the fruits of one's labors an education can be obtained concerning the forces at work in the bio-sphere. This book is intended as a signpost pointing to the reapplication of that knowledge in today's gardens.

The stimulus for most of the ideas in this book springs from the work of Dr. Rudolf Steiner and later practitioners of the bio-dynamic method. Rudolf Steiner's immense knowledge and practical understanding of the Earth Logos in today's world mark a spiritual grounding for modern man in this new age of the spirit. Man and plants must once again learn to co-operate in order for life to be supported. In terms of bio-dynamic gardening, Aristotle's view of the four natural elements may be seen as a foundation for the work of Rudolf Steiner. The resulting knowledge can then be applied to the task of practical experiments with the earth. The alchemical principles of Paracelsus–*Sal* (salt), *Mercur* (mercury), and *Sulf* (sulfur)–are ways of describing the processes at work in the bio-sphere among the elemental forces of earth/water/air/fire. These elements and elemental processes are the forces behind physical forms.

Chemists, physicists and agriculturalists share this space in common.

Through studying nature, the Greeks developed the science of phenomenal forces. Later on, Goethe, through brilliant observations, described the various plants of the world as a unified whole, by seeing the basic "plant" in the leaf. The phenomena of metamorphosis that Goethe described opened the door for modern man to enter into a conscious interaction with the plant kingdoms. Work of this kind has been undertaken by bio-dynamic investigators like Ehrenfried Pfeiffer and Maria Thun, who use the work of Dr. Steiner as a way of approaching the thought processes of the Earth Logos. Much positive information concerning the earth and the cosmos has been discovered and worked out scientifically by the biodynamic researchers.

The gardening year is a series of activities that must be carried out in a logical, rhythmic progression in order to be effective. The intention of this book is to present, both artistically and scientifically, some fundamental principles of the Earth Logos throughout one year of gardening activity. Although it is not an ephemeris or a calendar, it is meant to complement a publication such as the ***Stella Natura Calendar***. If read slowly and at the proper times, the verse can accompany the changes of the seasons and understanding will unfold in the gardener's soul like a seed in the soil.

The book is in four parts, the first chapter being a summary which concentrates a year's work into a few pages. This is for the overview so necessary in grasping the wholeness in gardening. Chapters two through four describe, in more philosophical terms, the alchemy of the Earth Logos and man's relationship to the soil and plants. The third section is composed of verses which describe the interaction of earthly and cosmic phenomena within our gardens, meant to stir the musings of the gardener's mind, and the fourth section is composed of a series of verses for the signs of the zodiac. The mandala drawings are designed to help clarify and develop principles which are difficult for words to explain. A leisurely reading during the various seasons will give the poetic Muse time to develop our understanding.

That I have borrowed heavily from others' work is due to the esteem I have for the value of the material I used in my research. For further reading, consult the bibliography at the end of this book.

Dennis Klocek

A Biodynamic Book of Moons

Chapter I: The Gardener's Year

In order to promote the ongoing of each species of plant and to safeguard sufficient propagation, Nature produces and germinates many more seeds than could possibly grow into mature plants. Many seedlings die as a result of crowding and adverse weather conditions, and as food for insects, animals and man. The wise gardener will follow Nature's example when sowing seed. Thinning seedlings is an art that helps the strong survive and provides much food for the gardener. If crops such as carrots, beets, turnips, and chinese cabbage are well sown, thinnings can be used for soups as they start to compete with each other for space. Once they are thinned to the proper space, they are left to mature into full-size vegetables. As seedlings are taken out, the plants left in the ground quickly fill the space so that, a day later, it is often difficult to tell which rows were thinned. Vegetables such as leeks, lettuce, cabbage and kale can be sown thickly in baby beds and moved out as selected plants mature. The remaining baby plants are then thinned and left to grow further, or taken for the table. In this way the baby beds always keep filling from one sowing. Vegetables too small for salad are excellent candidates for drying and storing for the winter soup pot; leeks, carrots and young greens are especially good for this purpose. Thinning works along with Nature's best economic principles.

As plants mature beyond the seedling stage, they exhibit specific needs with regards to space. Plants not given the proper space develop rot and disease and attract insect pests. A rule of thumb is that each plant should touch the leaves of its neighbor at maturity. This is especially true with heavy-feeding crops and most crops that are transplanted. Beans and peas do not mind being crowded and can easily be sown between the slower-growing crops, where they set seed quickly and die back. Beans can be sown between the vegetable rows at transplanting time and both crops will do well together.

The specific space requirement for each vegetable is usually found on the seed packet. When transplanting, try to visualize the mature plant in the row. Plants having upright growing tendencies such as onions, leeks, beets, turnips, celery and some lettuces can be sown or transplanted between larger, more space-demanding crops wherever space allows.

An old adage states that five acres of land well-manured and well-seeded will out-produce ten acres of poorly manured, thinly seeded land. The prime requirement for success is the diligence of the gardener in providing each plant with the proper amount of space. The ideal spacing in the vegetable garden would have a living mulch of vegetables over the entire gardening area at all times. Where there is no ground cover, the earth will produce a weed to cover her naked bosom. If instead of a weed the gardener cultivates a plant for food, this will grow and suppress the growth of the weed plants. Radishes, turnips, beans and lettuce work well as living mulch. Starting early in the year, you may sow peas in the spaces intended for summer squash or late corn. Onions sown early from sets, with a lettuce companion, mature at midsummer to be followed by kale or cabbage in the fall. The living mulch crop is sown as companion to the early crops just mentioned, and the short season allows them to mature before the early crop comes out. Lettuce and onions like to share space; radishes, peas and lettuce are wonderful companions. Leeks do well sown between the beet rows. Slow-to-germinate crops like parsnips, carrots and parsley can be sown with radish or early turnip seed in the same row. The radishes help mark the rows for cultivation and are eaten long before the other crops mature. The American Indians traditionally grow the "Three Sisters" of squash, beans and corn together in the same field. A hill of corn alternates with a hill of squash, with dry beans filling the spaces in between. Pole beans are sent up the corn stalks on the outside of the plot.

Summer peas are the traditional legume grown under corn in the farms of the South. Care and knowledge are needed to get the spacings correct. In their natural state, however, plants grow gregariously, giving support to each other. It seems that the earth just wants to cover herself with green.

In order to give direction to the complex activities of intensive gardening it is advantageous to establish permanent garden beds. The accepted dimensions for each bed are approximately four and a half feet wide by twenty feet long. Early in the spring, snow melts on the south and west sides of the raised beds much more rapidly than on level ground. Runoff water stays in the paths, making the beds accessible much sooner and retaining much of the runoff. Raised beds encourage air circulation and put the roots of the plants into a zone of increased etheric activity. Established beds become the gathering and spawning place for large quantities of earthworms. They leave their fertile, balanced castings all over the beds during the entire winter, if a protective mulch is put down in the fall. Permanent beds with well-defined pathways allow children, animals and adults to enter the garden and enjoy the plants without compacting the soil of the growing area or damaging seedlings. Small children can be easily taught to follow the path. A garden marked off in permanent, twenty-foot beds seems to gain vitality from the very act we undertake in mounding the earth. From the neolithic Seip mounds of North America, to the sophistication of the old and new world pyramids, mounds have long been a recognized method for attracting and amplifying basic earth energy. The raised and mounded shape of the mound concentrates and enhances the etheric activity of Nature's processes.

If a mound is composed of the remains of plants and animals mixed with soil and kitchen wastes, the concentrated earth energy converts these into humus. This mound is then a compost heap.

The earth herself is too dense and dead for the plants to assimilate into their organs. The plants must rely on an intermediary for their food, even as we rely on them for our nourishment. This intermediary is humus, which is part plant (colloid) and part mineral (soil). The compost heap is the perfect mound shape needed to concentrate earth energy into the formation of humus. Materials piled into a mound will compost three times faster than materials that are dug into the soil or merely left unmounded. Raised garden beds and compost mounds capture and intensify energy for the organic processes. To build a heap from the ground up, cut a circle out of sod. Pile the sods around the rim of the circle. Turn them upside down and sprinkle the roots with lime to help them rot. Within the circle, blend the materials being added like a cook. Add kitchen wastes, leaves, grass clippings, and manure as they come. Avoid heavy concentrations of meat in order to keep the animals from disturbing the pile. Keep the mound moist but not soggy and build it in the shade of some birch or alder trees, over an old nettle patch or where a heap or manure yard has stood. Coffee grounds, cottonseed meal or old cornmeal, sprinkled on the ground before you build the heap, attract the earthworms. Wait for rainy periods before adding dry materials in quantity. If too much rain makes the heap too sodden and waterlogged, it will start to have a bad smell. It is easiest to ventilate the heap by poking holes in it with a locust pole sharpened to a point. Air will percolate through eighteen inches of compost to sweeten the sour reaction and ventilate any excess heat. If smells persist, drop slaked or hydrated lime in each hole and drive in some more ventilation shafts. After heat leaves the heap, put in the bio-dynamic herbal preparations.* These intensify the action of the heap, help it retain plant nutrients and vitality, and help improve the health and condition of the soil. When the heap has shrunk by half and is dark and sweet, it may be used for planting seed or transplanting without burning the plant.

The science of crop rotation is called agronomy. It is the study of which crops are best suited to follow other crops on the same piece of land. As well as taking nutrients from the soil, crops use up specific earth energy for the formation of their organs. In order to restore balance to the soil, the farmers rotate crops that develop the same organ consecutively in the same field. A typical grain rotation in a large field is corn, wheat, and alfalfa. After

** See reference list at end of book.*

fertilizing in the fall and letting the fertilizer work into the soil for the winter, a heavy-feeding crop such as corn is sown. The corn is then harvested in the summer or fall and wheat is sown to winter over. Early in the spring the wheat is turned under as a green manure or the grain is harvested in the summer. After the wheat harvest, the field is sown to alfalfa. This is cut as many as four times a year, for a period of up to five years, and used as hay for animals. Being a legume, the alfalfa feeds the soil and improves its structure. When heavy-feeding crops are again desired on the field, the alfalfa is turned under, fertilizer is put down and the rotation is begun again.

Farmers strive to keep their rotations regular in order to keep up with the weeds in the larger fields.

The gardener has an advantage over the farmer in his ability to improve the soil to a much higher degree than is possible in the fields. Intercropping allows two or more rotations to happen simultaneously in the enriched garden soil. Cultivations are reduced as a result of intercropped vegetables acting as a weed-suppressing mulch. Intensive gardening practices are most efficient with the use of hand tools, thus eliminating large capital investments and the depreciation of expensive machinery.

Intensive farming requires that strong efforts be made to improve the health of the soil to the maximum with compost. This reduces the need for fertilizers and eliminates the need for pesticides. Intensive gardening requires that each plant be given the proper nutrient, time, and space to grow into a mature, disease-free plant of high food quality.

An essential aspect of agronomy is the fact that the growing season is much longer than the summer months. By observing the changes in the large fields of winter grain, the gardener can get an insight into the rhythms of nature, and extend the gardening season. As the sun passes before the constellation of **Pisces in March/April**, the watery, fruitful character of Pisces causes the brown grain plants in the fields to be touched with a subtle green. This delicate green signals the beginning of the planting season. The sun has been rising for two months and the earth is unfolding her powers of germination. At this time gardeners can begin planting the cold-resistant crops, months before the farmer can venture onto a field with heavy machinery. Biennial seeds such as parsnip, carrot, leek, beet, and chard benefit greatly by being planted early in Pisces. Crops that are grown for their roots need a lot of time to mature and sweeten. Roots intended for winter storage should also be planted at this early date: these include potatoes, late carrots, winter-keeper beets and all late roots in general. This is the time of strong root development. Crops such as peas, onions, lettuce, spinach, mustard greens, and early turnips thrive in the cold spring rains and bolt to seed or become bitter at the first sign of heat. All of these crops are frost-tolerant and can even stand some freezing in the soil after they have germinated. As the sun moves before **Aries in April/May**, the grains in the fields will turn to a dark green and start to rise on their stalks. This dry, barren time is a signal to hoe and cultivate all the seedlings that germinated in Pisces. Hoeing at this time is very effective against the incredible rush of spring weeds that can otherwise overtake the small plants. Hoe the leeks, beets, carrots and onions at ten-day intervals, each time the moon is in front of an earth sign. After hoeing for a month and a half, mulch well and plant wax, dry or bush green beans between the rows. The beans push up through the mulch, suppress the growth of weeds and provide beans in Taurus.

When the grains in the fields start to form heads, this tells us that the earthy, **Taurus (May/June)** sun is quickening the earth's fertility. The heat-loving fruit and seed crops can now be sown or transplanted out into the garden. Corn, squash, tomatoes, peppers, eggplant, cucumbers, and melons of all kinds are quick to germinate in Taurus earthiness, and as summer approaches they find ideal conditions for rapid growth.

At the summer solstice, in airy **Gemini (June/July)**, the winter grains have set seed and are beginning to ripen. As the sun starts to decline, there is a rest in plant development. Strength is gathered for the all-important fruiting and seeding of the summer. Hoeing, mulching, and harvesting occupy most of the gardener's time.

After the solstice rest, planting is again resumed for the fall crops like cabbage, kale, broccoli, chinese cabbage, cauliflower, and collards. Seeds sown late in Gemini and transplanted out in watery Cancer grow strong as the "second spring" growth spurt vitalizes them in **Leo (Aug./Sept.)** and in **Virgo (Sept./Oct.)**. These crops can take the place of peas, onions, and lettuce, when they die back at midsummer, making the garden lush and productive as frost blackens the more tender crops. Compost added to the transplanting hole speeds the growth of the leaf crops into the fall. The gardening year between **Pisces (March/April)** and **Leo (Aug./Sept.)** is devoted to the annual plants and the food crops that are relatively quick-growing. With the onset of Leo, however, another gardening season commences that is devoted primarily to plants in their dormant stage and to the woody, long-lived trees and shrubs. It is divine providence that has organized the gardener's year with the proper time to turn to each task. When Leo is in the summer sky, cuttings are taken from the half-ripe wood of the evergreens. These are heeled in under lath and later under straw and kept moist and cool for a whole year. Bramble fruits are layered now, and next spring young plants will shoot. (Fig. 1)

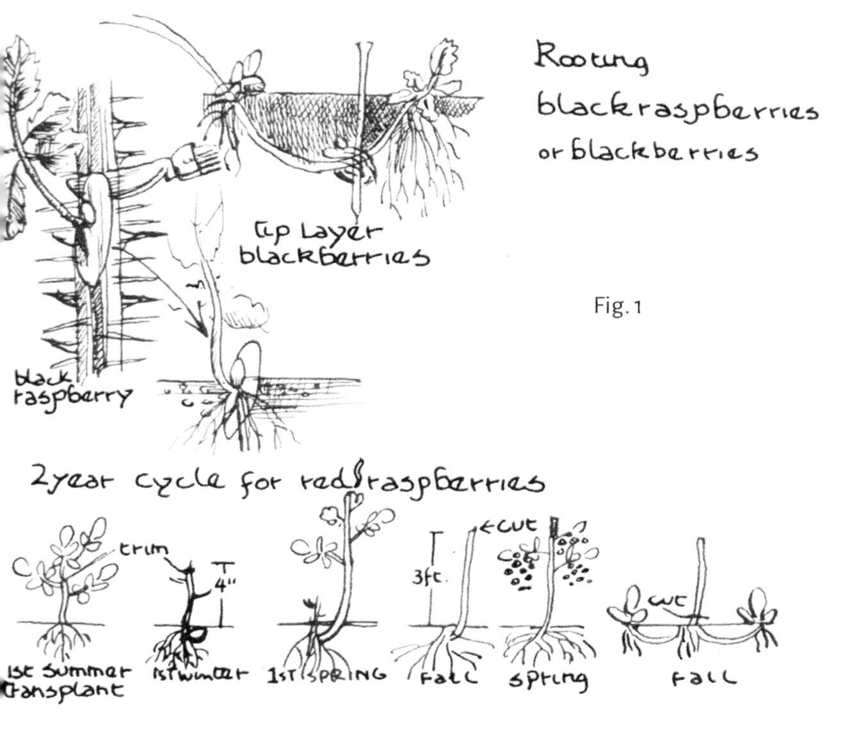

Fig. 1

As **Virgo (Sept./Oct.)** graces the clear, fall, daytime skies, half-ripe cuttings can still be taken. (Fig. 2)

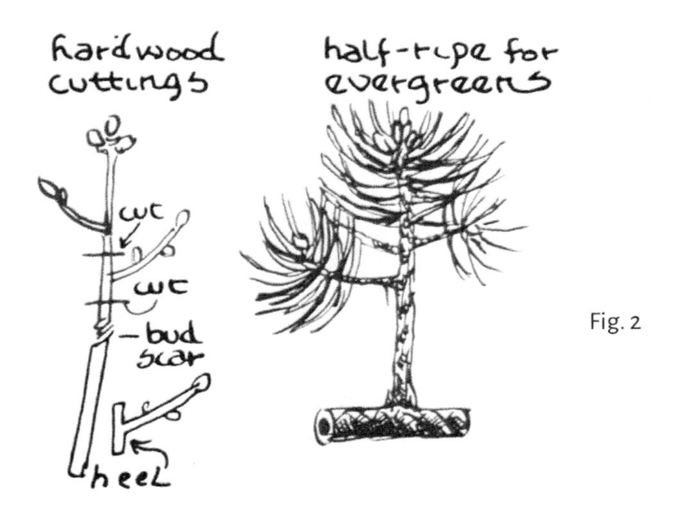

Fig. 2

Clumping roots of plants and herbs can now be divided and long-lived trees and bushes safely transplanted.

Seed for fruits can be sown directly from the fruit into the cold, moist ground of Virgo. These will germinate next spring and serve as rootstocks for grafting a year from germination. (Fig.3)

Libra is a constellation of short duration. The dry barrenness of Libra is an aid to the storing and harvesting of the winter food crops. Mulch crops against the winter cold in **Libra (Oct./Nov.)**. As **Scorpio (Nov./Dec.)** arrives, the trees have ripened off their

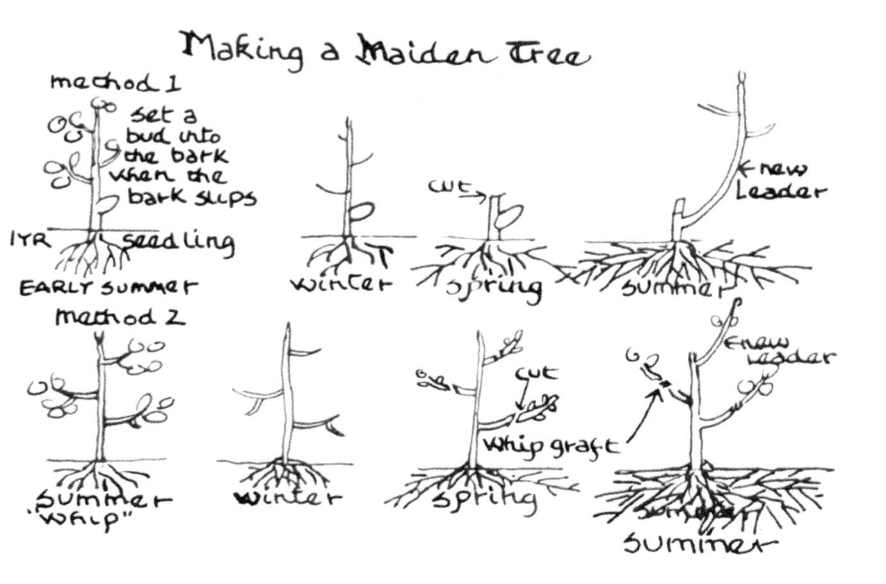

Fig. 3

growth and the wood is hard and dormant. Cuttings taken with a heel are stored through the winter and rooted in compost early in Pisces. Sagittarius is a barren fire sign and marks the point of lowest descent of the sun at the winter solstice. **Capricorn (Dec./Jan.)** is an earth sign and marks the point when the mineral body of the earth is most sensitive to the cosmos. The crystal forms of ice and snow are examples of this increased mineralization. **Aquarius (Jan/Feb.)** is an air sign ushering in the intense, life-renewing cold of the outer planets and bringing out the light from the depths of winter.

With the onset of **Pisces (Feb./March)**, the planting season once again commences. Grafts are set and scions are cut as the watery, fruitful character of Pisces rekindles the fertility of Mother Earth. (Fig. 4) Pisces is the proper time for setting out wintered-over hardwood cuttings. The roots will be encouraged by the ample moisture. After the spring equinox, plant life is quickened.

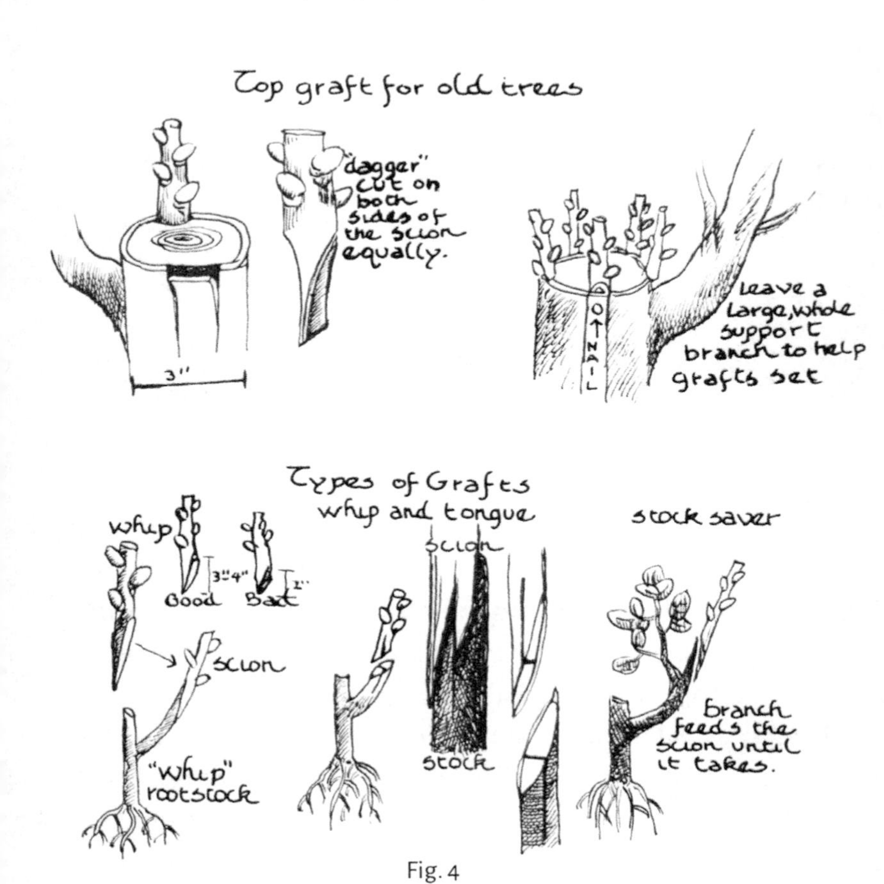

Fig. 4

Aries is the best time for safely pruning fruit trees. The trees are alive in the cambium and the cuts heal quickly in the dryness of **Aries (Mar./Apr.)**. (Fig.5)

Fig. 5

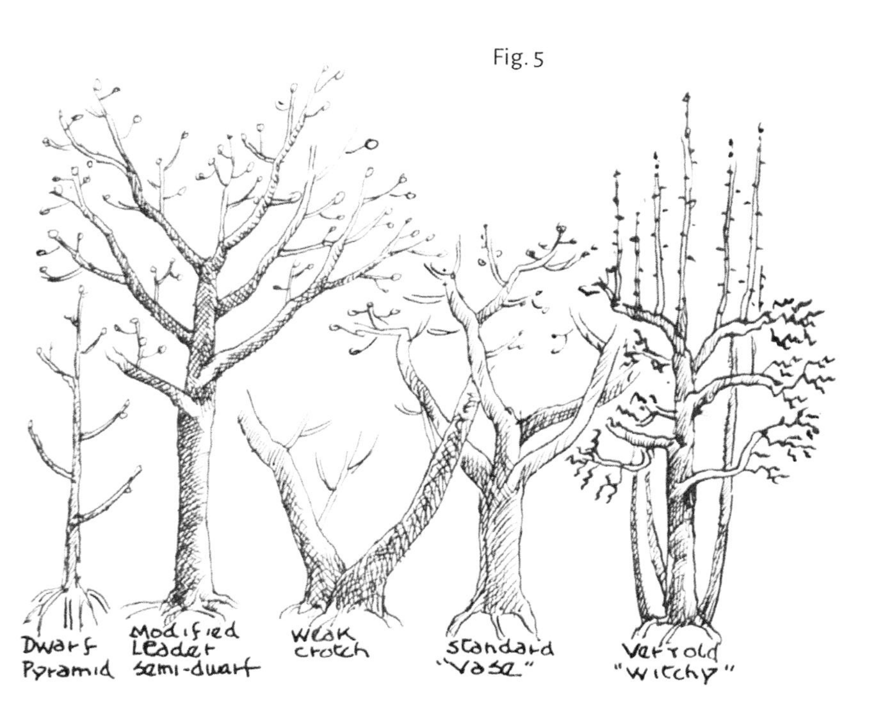

Taurus (May/June) marks the shooting of the new wood. Softwood cuttings may be taken now, while the new wood snaps. They will root quickly in the earthiness of Taurus, to be transplanted out in **Cancer (July/Aug.)** or **Virgo (Sept./Oct.)**. Taurus also marks the slipping of the bark on the trees. This event signals the end of the season for setting grafts with scions, and the beginning of the bud-setting season. (Fig. 6) This method places the buds or "eyes" from the parent branch under the loosened bark of the rootstock for propagation. Buds may be set during the entire summer season until the bark gets tight in the fall.

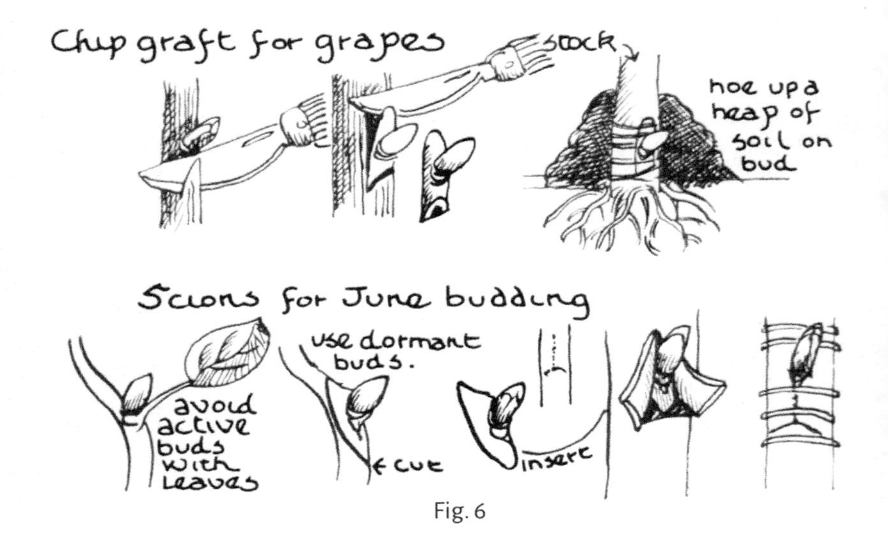

Fig. 6

Thus an entire year can occupy the gardener within the elemental rhythms of the stars in the Zodiac. Although conditions in any given year may delay or accelerate the time periods by as much as a month's difference, Nature's wholeness balances these differences so that general planting guidelines can be developed.

Potatoes, parsnips, late carrots, beets, late leeks and chard are planted early in Pisces and occupy the same garden space for a whole year. These are the root crops. Squash, tomatoes, peppers, eggplant and melons are planted indoors in Pisces and moved out in Taurus. Corn, squash and pumpkins are best sown in the ground in Taurus. All of these crops are fruits and occupy the same garden space for the whole year.

Radishes, peas, onions, turnips, mustards, lettuce, beets and wax or bush beans are either planted early and harvested at midsummer or as they mature, or planted at midsummer to mature in the fall. They occupy the same garden space for half a year.

Lettuce, cabbage, kale, broccoli, cauliflower, brussels sprouts, and chinese cabbage are easily sown in late summer and transplanted into rows as the garden thins out in the fall. These are the cool-weather leaf crops.

Wax beans and green bush beans may be sown at any time after danger of frost is past. These grow for half a year and die off. If planted at midsummer, these beans can be eaten in the fall.

Lima beans, black beans, pinto, soldier, pole, aduki, and turtle beans are all dry beans. These are sown in Taurus after frost and occupy the same garden space for the whole year. Fava beans are also dry but are sown in Pisces. Parsley, parsnips, carrots, and kale and collards, chicory, salsify roots, leeks, garlic, and shallots can survive a cold winter in the ground if mulched after the ground freezes.

Endive, winter radish, lutz winter beets, burdock, and rutabaga may be stored in damp sand away from frost for winter eating.

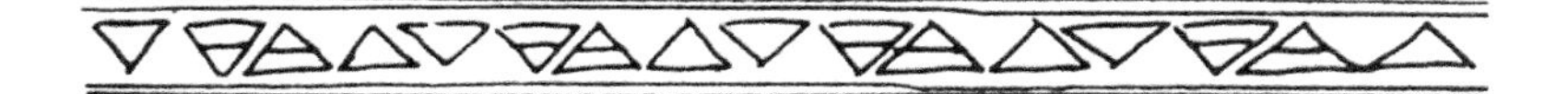

Chapter II: Processes in the Earth Body

The earth is a living being. Just as her creations must be born and die away, she too must go through a continual transformation of elemental substances. These transformations are living examples of her gesture as a being. In this way she continually builds bodies for new creatures out of the discarded bodies of those that have moved on. These transformations are dependent upon three processes used by alchemists since Paracelsus to describe the miracle of life resurrecting from the remains of the dead. The three processes are ***Sal*** (salt, absorption, precipitation, coming into being), ***Mercur*** (catalytic change, mercury repetition) and ***Sulf*** (sulfur, burning, dissipation, fragrance, sublimation). (Fig. 7)

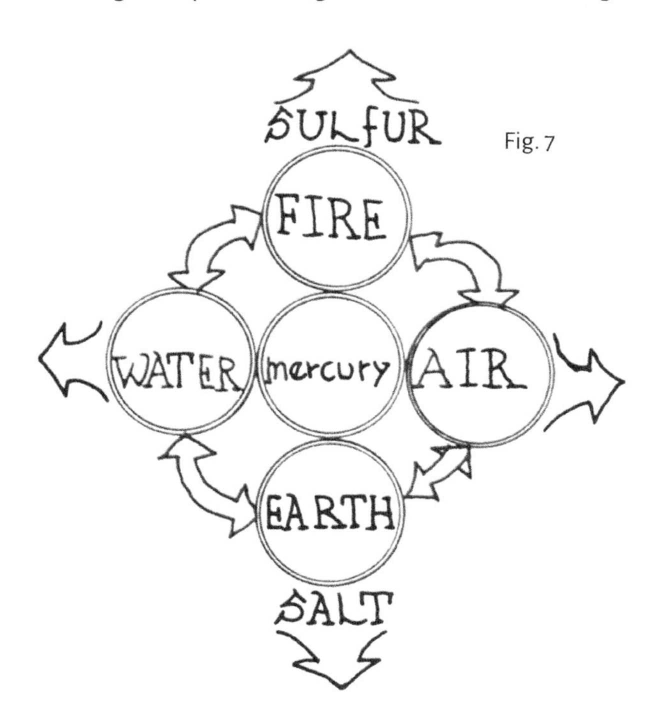

Fig. 7

In the mineral kingdom the ***Sal*** processes are those which tend toward the limestone pole. Through the absorbing (***Sal***) power of the lime the earth precipitates (***Sal***) the root out of minerals.

Spreading, branching roots, typical of most limestone-lovers, show a strong ***Sal***/absorptive function. The absorptive (***Sal***) quality of the legumes is intensified to the degree that they precipitate (***Sal***) nitrogen from the atmosphere into the soil.

The opposite process of a ***Sal*** (becoming) process is a ***Sulf*** (dissipating) process. Burning is characteristic of the ***Sulf*** (sublimating) activity. In the ***Sulf*** process, the dross is burned out of the compound through the action of fire. The result is ash from a ***Sulf*** reaction.

In the mineral kingdom, silica/phosphorus is the ***Sulf*** (burning) pole. The crystals of light-filled quartz and fuming phosphorus represent the power of the sun among the minerals. (Fig. 8)

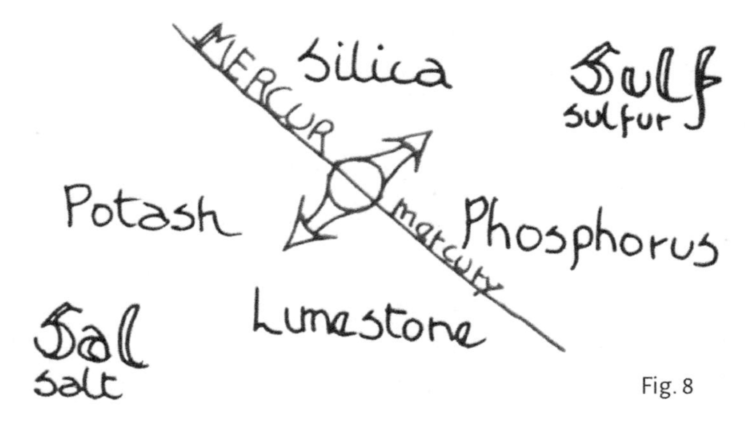

Fig. 8

As a mineral, quartz is the parent of all rocks. Most granite, gneiss, mica and some shale and sandstone come from quartz beginnings. It is therefore continually giving of itself to develop other rocks. In like manner the great sun is continually giving of its material body So that we may live. Both the sun and quartz are dissipating (***Sulf***) themselves for a higher purpose.

Phenomena of ***Sal*** and ***Sulf*** are most easily observed in simple reactions like salt in water. Salt (***Sal***) crystals have a strong ability to absorb water. When in complete solution, the salt is in a ***Mercur*** (catalytic change) relationship with the water. The water supports change without changing itself. This and its flowing character make it mercurious. Conversely, when the element of heat is applied, the ***Mercur*** leaves the solution and the salt again has its original form precipitated (***Sal***) out of solution.

An opposite process is the burning of a stick. (Fig. 9) This is earth being consumed by fire, giving up ***Sal*** and ***Mercur*** by releasing water and gas. Instead of remaining as the salt in a ***Sal*** process, the earth (wood) is being sublimated (***Sulf***) to produce light and heat. The dregs are burnt out of the compound, producing acrid smells when sublimation is partial and fragrance when sublimation is complete.

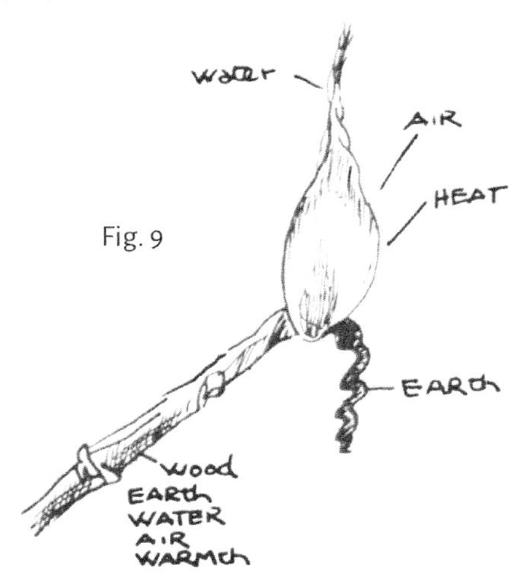

Fig. 9

In this instance the compound is unable to gain its former appearance. The result is that water and earth go in separate directions as gas and ash. They are sublimated and released from bondage by the intensity of the fire.

In the plant kingdom the ***Sulf*** process is most clearly seen in the plant that has reached flowering. The ***Sulf*** (dissipating, sublimating) process of the lower root/leaf plant, and the metamorphosis of organs from gross, water-filled leaves to fine, light-filled membranes, represent a true burning. (Fig.10) The plant is literally burning at the expense of its lower nature in order to "parent" seeds which will go beyond.

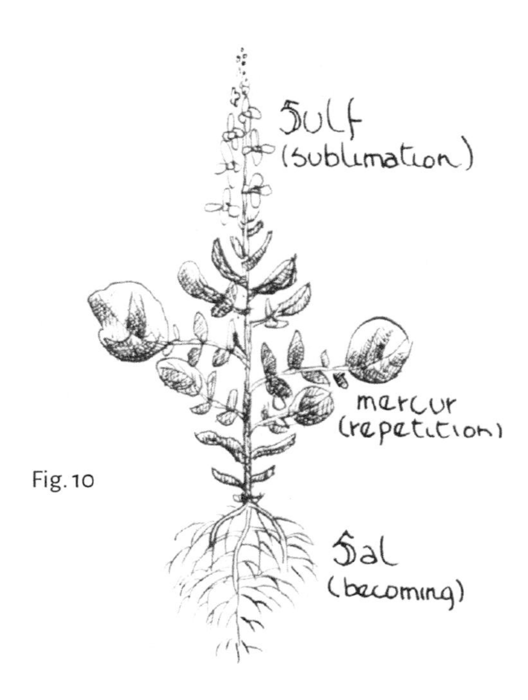

Fig. 10

In contrast to this dissipation (***Sulf***), the emergence of the root from the seed is a sure sign of the ***Sal*** (coming into being) of the new plant. The entire process of germination is extremely ***Sal*** (absorptive) with regards to water and earth. (Fig.11)

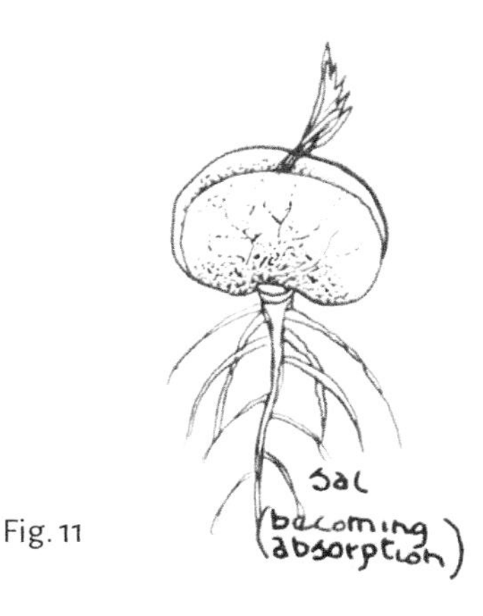

Fig. 11

Sal (absorbing, precipitating) processes govern the root/ leaf formation in general ***Sulf*** (burning, sublimation, fragrance) processes are generally found in the flower to fruit stages. Each extreme has a tendency to move away from its opposite. In order to cause interaction, a third set of variables is required. The ***Mercur*** process (change, repetition, catalysis) is the great harmonizer between the extremes of ***Sal*** and ***Sulf***.

In the mineral kingdom, clay is the great neutralizing catalyst effecting force changes through water, taking it on and passing it off, physically remaining always neutral between silica and lime. These forces are found in the leaf systems of plants. In leaves the repetitious character of ***Mercur*** processes reaches incredible dimensions. The ***Mercur*** process is found in the flower petals, anthers, stamens, and seeds of the flowering plant, but here it is not working alone, for the flowering process is an intensified microcosmic event for ***Sal***, ***Mercur*** and ***Sulf***.

The plant springing from its roots is strongly ***Sal*** (becoming). The repetition of leaf after leaf is ***Mercur*** (change). The plant moves into a contraction and begins to sublimate (***Sulf***) its physical body to produce the flower. (Fig.12)

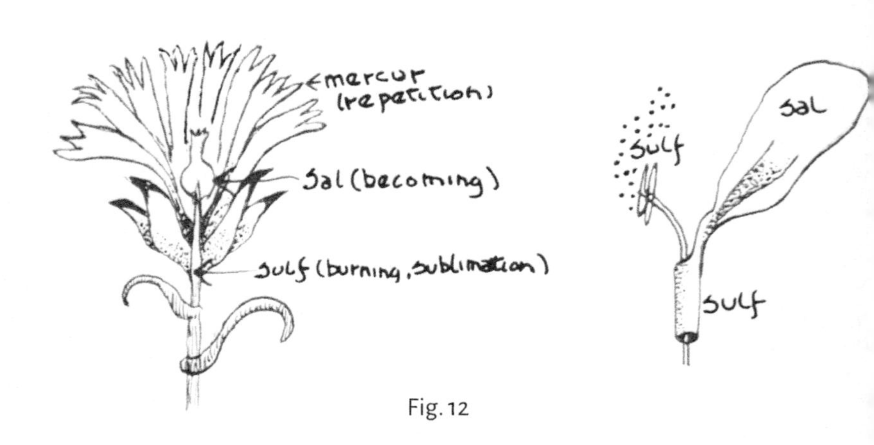

Fig. 12

Nodes become smaller and leaves move closer until all growth centers around a central point. This point is often the center of the calyx. The image is of a sharp silica needle tapering into nothing. At this point the ***Sulf*** process takes over the development of the life energy of the plant. But nature requires balance, and the contracted plant soon expands and produces flower petals in repetition (***Mercur***). The ***Sal*** process is also part of corolla formation. As the petals come into being, pollen (***Sal***) is precipitated out of sap. But simultaneously strong ***Sulf*** (fragrance, sublimation) forces are contracting petals into thin tubes of filaments and producing fragrance (*Sulf*) as the flowering plant sacrifices itself for the seed. The intense *Sulf* activity of fertilization reaches another point in the seed, and this initiates a *Sal* (coming into being) reaction as the fruit swells by precipitating (*Sal*) nutritive substance from plant sap. The ramified veins of many fruits represent a *Sal* (coming into being) gesture from the potash and can be traced back through the leaf to the root.

In a final gesture the plant destroys the fruit and ripens the seed in a *Sulf* reaction akin to burning. This final contraction locks into the seed power of a fiery, cosmic, chaotic character. To germinate, the intense *Sulf* process in the seed must contact water, which reintroduces the *Sal* properties of absorption. The seedling then dissolves the store of food in a *Sal* reaction where the root is discharged and a new plant is brought into being. The alternation of *Sal* and *Sulf* exactly describes the action of potassium and phosphorus in the life of plants. (Fig. 13)

Fig. 13

Potassium (*Sal*) develops root and stem, aids fruit production and seed viability. Phosphorus advances ripening in fruit, stimulates flowering, and develops prolific seeding. The gas, nitrogen, has a decidedly mercurious function in plants, as it opens up the cells to produce swelling growth. Being a gas, the nitrogen is ever ready to escape with methane. The rapid changeability of nitrogen makes it a perfect catalyst (*Mercur*) that can bring potassium and phosphorus into fruitful relationship with each other.

potash - nitrogen - phosphorus
Sal - Mercur - Sulf

Plants that produce large leaves or strong, rank growth all have a great need for the mercurious character of nitrogen. By tenderizing and opening up plant protein through strong catalytic (*Mercur*) chemistry, nitrogen also increases the speed of growth. This accelerated growth works best in the vegetative parts of the plant, i.e. where leaf follows leaf with little or no metamorphosis. If excess *Mercur* properties are given to the seedling through excessive nitrogen fertilizers, the flowering is delayed because the *Mercur* (repetition) nitrogen must repeat leaves and the *Sulf* (sublimation) process is retarded. If the *Mercur* process becomes entirely too rampant, then the changing catalytic energy causes the cells to open too much and fungus, rotting stems or insect attacks are a result. This type of destruction represents a *Sulf* (burning-rotting) reaction to an excessive *Mercur* process. Since the abundance of nitrogen will not allow the plant to reach flowering, the flowering (*Sulf*/ sublimation) process moves down into an area where there are no organs for flowering. Fungus and insects can be considered as substitute flowers, from the plant's point of view. From the gardener's point of view, they are mirrors of the imbalance between *Sal*, ***Mercur*** and ***Sulf*** processes in the soil.

Thus the nitrogen, phosphorus, and potassium of today's agriculture can be seen in holistic relationships with each other instead of simply as "plant nutrients" to be provided for larger produce. But these relationships do not stop at the level of plants and soil In the animal kingdom, ***Sal*** and ***Sulf*** and ***Mercur*** can be seen in most life functions.

For instance, in respiration the inbreath which absorbs oxygen is ***Sal*** (absorbing/becoming). The oxygen is absorbed into the blood and CO_2 is precipitated (***Sal***) out. ***Mercur*** is found in the repetition of inbreath and outbreath, and the ***Sulf*** process is found in the outbreath which dissipates (***Sulf***) carbon dioxide into the atmosphere.

A further development of the ***Sulf*** process is in the "burning" of the food with the oxygen in the process of metabolism. Sublimation (***Sulf***) occurs when gross nutrients are changed into energy. The ***Sal*** process is also active in metabolism, precipitating body substance from body fluids. Transmutation of one substance into another is brought about by the subtle interaction of these processes. In the natural world, transmutation provides the mysterious link between plants and mankind. Plant sap and human blood are remarkably similar substances, except that magnesium and chlorophyll are replaced with iron and hemoglobin. Due to this unique transmutation, men and plants are at opposite metabolic poles. Plants transmute iron in the soil and absorb CO_2 to produce magnesium, the basis of chlorophyll.

(Fig.14) Men transmute the magnesium of plant cells into iron and absorb o2 to produce iron (blood).

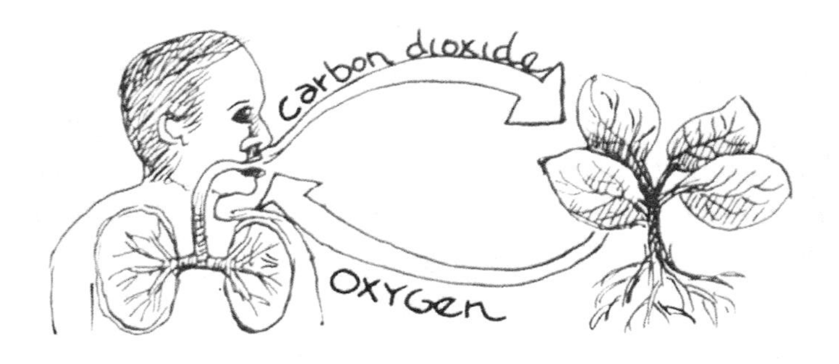

Fig. 14

The three processes of *Sal*, *Mercur* and *Sulf* provide much insight into the alchemical view of creation in nature. Any creature living in the world must first be born out of nothing (*Sal*). The absorptive qualities of the *Sal* process underlie the formation of the physical body, especially in the formation and function of nerves, brain tissue, lymph and blood. The thought processes so dependent on potassium in the brain are absorptive in their function, and knowledge is precipitated in the form of concepts by the thought processes. Nerves transmit impulses through *Sal* reactions between the two salts, sodium and potassium. Blood and lymph also balance sodium and potassium, allowing cells to pick up nutrients through absorption and to pass off waste through the cell wall by absorption in the opposite direction. *Mercur* elements in organic chemistry are nitrogen, hydrogen, and oxygen which unite with *Mercur* carbon to form energy-producing sugars and fats.(Fig. 15)

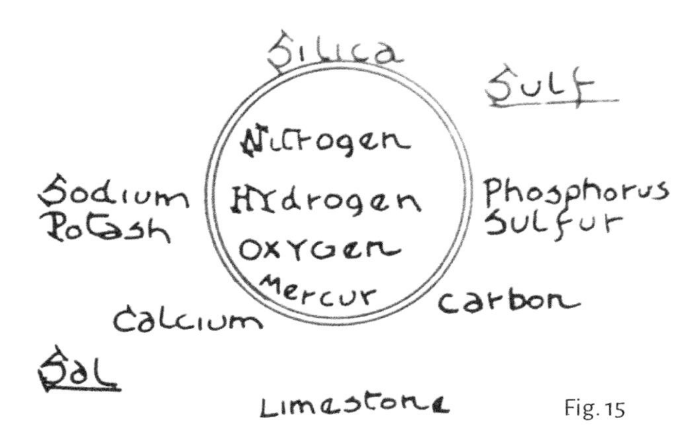

Fig. 15

Carbon and silica are minerals that possess the property of easily forming with oxygen into acidic compounds. Most elastic tissues like muscles and tendons are largely protein, composed of carbon, oxygen, hydrogen and nitrogen, all very *Mercur*. The expansive and contractive function of the *Mercur* aceous muscles is purely repetitious and catalytic, i.e. always changing but always staying the same. Waste products are easily combined with oxygen, and gases (*Mercur*) are passed off to be absorbed by the blood. This middle system of respiration/circulation is rhythmically *Mercur*. (Fig. 16)

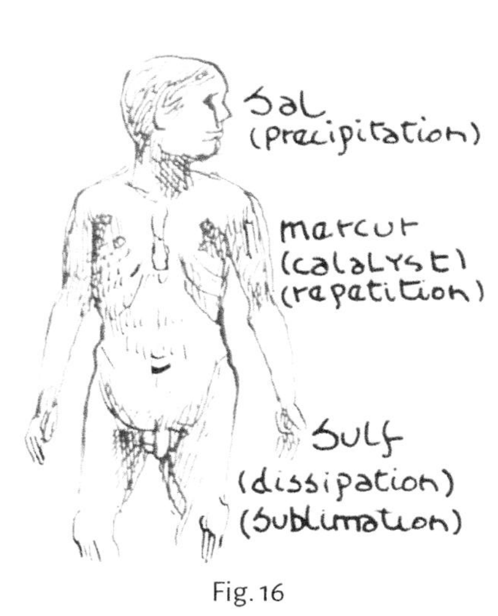

Fig. 16

The third process (*Sulf*) is the fuel process wherein fats and carbohydrates are destroyed by the metabolism to create energy or body substance.

Breakdown of fats and proteins through phosphorus and sulfur produces energy plus *Mercur* gases: nitrogen, carbon dioxide and oxygen. The metabolic process has various phases. The initial phase is the breakdown of complex compounds. This is a burning phase. The second phase is the elimination of waste. This involves *Sal* reactions as well as *Sulf* deposits of high concentrations of phosphorus in urine and solid waste. The third *Sulf* process is the building of body tissue. Here the *Sulf* processes produce sublimation by using the broken compounds of carbon, oxygen, nitrogen, potash and phosphorus with hydrogen to reconstruct proteins, nerves and bones. The new nerves and bones are the last stop in the metabolic process. An imbalance in the food intake is deposited in the bones as a sort of ash. Too much phosphorus makes the bones brittle and hard, with their life burned out of them The lymph marrow is turned into hard bone in the typical *Sulf* reaction.

The fourth *Sulf* process is the ability of phosphorus to affect energy transfer. Biochemical synthesis alone does not explain how energy is given to the body. High phosphorus compounds are concentrated sources of sublimated solar energy. When breakdown occurs, this energy is released as fuel for the other life processes. The result is that one form of life is destroyed for the sake of another. Plants destroy their lower selves in order to produce proteins and oils rich in energy. These are eaten by animals and people who must destroy them in order to use them. The processes in plants of absorption, catalytic change, and sublimation are transferred to animals and serve the exact same requirements in their life functions. Thus energy and matter follow continuous patterns of forces as they move through body after body in nature's creations. These forces and processes are so coherent and whole that the blueprints for their interaction are found at every level of creation. In the life of an organism, for instance, birth is intensely *Sal* "coming into being." Once the being has incarnated and absorbed a physical body from its mother, it is subjected to the *Mercur* process working in the constantly changing parade of phenomena.

Upon aging, the organism begins to sacrifice the physical body back to the universe in a *Sulf* gesture of self-burning. If the organism is a plant or animal, this means being the physical food for a higher life form. In human beings, *Sulf* becomes the flowering of wisdom in the destruction of the body as the soul sublimates the life energies for the higher form of the spirit. At death the body is dispersed (*Sulf*), sending its elements into the earth to be used for her further creativity. The soul is then freed of *Sal* and *Sulf* and exists in the Eternal All of the Universe (*Total Mercur*).

Numerous examples of this process are given in the phenomena of life to act as a blueprint to the soul for the workings of universal evolution. Building a fire and cooking a meal can reveal the entire spectrum of possibilities contained within an earthly existence. The simple act of lighting a candle can reveal the alchemy of life. The solid earth and its *Sal* processes are seen in the crystallized wax. The *Mercur* processes of water and air are seen in the combustion of wax into a gas, the radiance of the light and the molten state of the wax. The *Sulf* process is developed by the fire element sublimating the lower form of the wax into warmth.

This view of the *Sal*, *Mercur* and *Sulf* processes needs to be woven into the structure of still another group of alchemical concepts in order to be satisfactory to the gardener. The concept of the four elements is the most ancient alchemical system and it is within this view that nature and man are one within the great plan of creation.

Chapter III: The Theory of the Four Elements

To find the roots of the chemical theory of agriculture it is necessary to look at the alchemical ideas of Paracelsus that are suggested by the processes of ***Sal***, ***Mercur***, and ***Sulf***. Originating in China, the theory of the four elements was already well-developed by the time of Aristotle. The method of scientific inquiry into the natural world was based on observation and intuition rather than on hypothesis and experimentation.

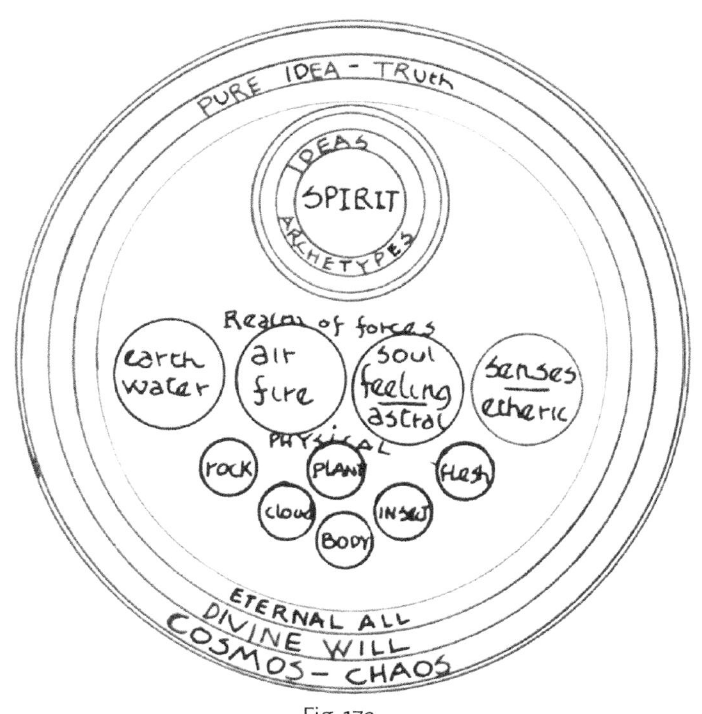

Fig. 17a

Aristotle believed that the physical world was a reflection of a world in which there existed pure, archetypal forms in a perfected state of development. What we see in the physical world is the end product of a series of metamorphoses starting with the pure idea or thought of the Creator and coming down to the earthly existence of creatures, plants, and minerals. (Fig. 17a)

These "ideas" or creative thoughts of the Creator of the Universe are then the source of the manifold energies and forms found in physical reality.

The "ideas" arrange themselves in physical reality according to patterns of behavior. These patterns are present in all living things to a greater or lesser degree, and are the result of the forces of Divine Will moving through the phenomena of the finite universe. The force patterns behave in such a characteristic manner that they are termed "elements" or fundamental, universal phenomena. Earth, water, air, and fire are therefore much more than the physical earth, water, air, and fire. They are qualities describing the subtle interplay of forces rather than a solid, physical composition or mass. Interaction between the four elements takes place in the realm of forces that guide all natural phenomena.

From a chemist's point of view this theory is naive and simplistic. The modern periodic table of elements is indeed a brilliant, complex insight into the atomistic world of compounds. In the chemical laboratory, Aristotle's four elements would prove to be woefully inadequate. What is needed there is precision and verifiability through experimentation. In the spheres of nature, and particularly in the labors of agriculturalists, the effectiveness of the chemist's periodic table is actually a hindrance. The gardener, as opposed to the chemist, must be aware of wide-ranging natural influences on plants and work within systems that have wholeness as the goal. It is in these whole relationships of mineral, rain, sun, plant and energy that the profound wisdom behind the theory of the four elements becomes manifest. Earth, water, air, and fire have specific influences on natural phenomena, and the understanding of their forces is a powerful tool in the garden. (Fig. 17b)

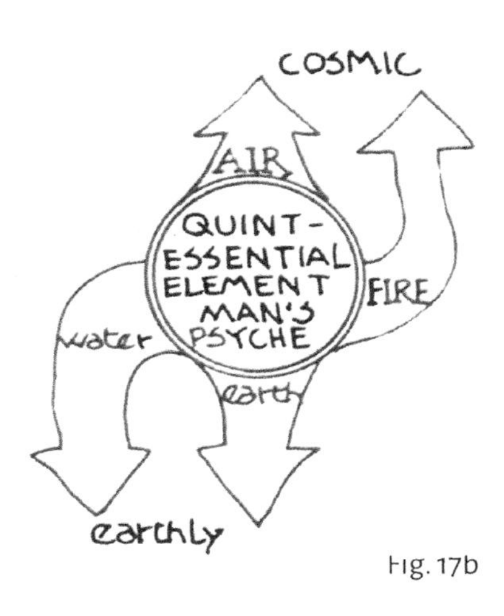

Fig. 17b

Unity in the natural spheres is a product of dynamic balance between extremes. Earth is heavy, dark and dense while its opposite, air, is light, bright and open. Water is cold, wet and moves downward, while fire is hot, dry and moves upward. The alternation of these elemental qualities and movements gives rise to local chemical reactions and global weather phenomena at the same time.

In the winter the crystal-forming power of the earth element predominates. The minerals of the earth body are intensely alive with the energy brought into the earth when the plant life returns to its roots. Water characteristically assumes the form of the intensified earth energy by producing ice or–when earth and water meet air–snow. As the sun moves toward the spring, the earth element begins to interact with the water element, and fogs and cold rains awaken the dormant plant life into growth. In the spring the water element predominates. Succulent, leafy growth of grasses, trees and herbs is strongly characteristic of water. As the sun moves toward the summer solstice, the element of air urges plant life to rise up from the earth in order to flower. Water leaving the earth and rising into the atmosphere is an image of the elemental activity of the hazy days of summer. With

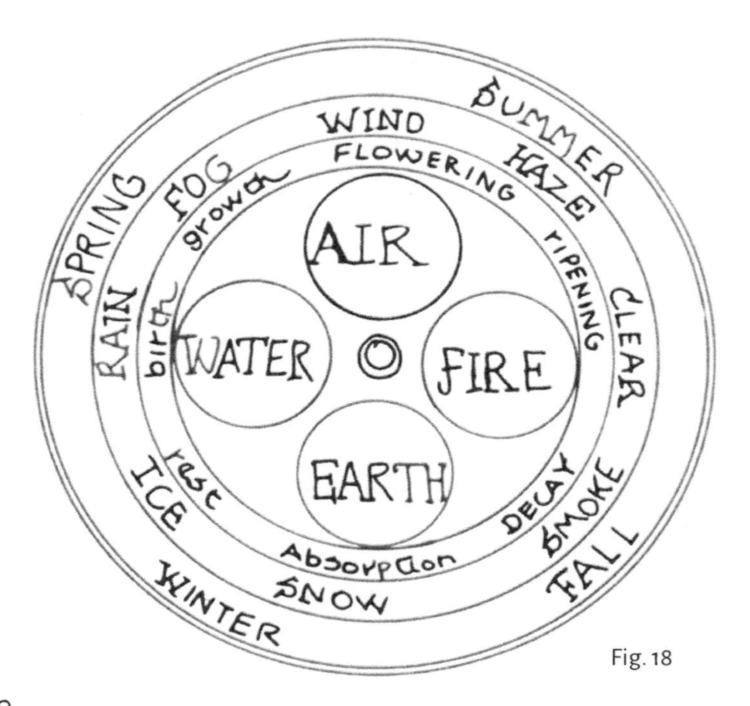

Fig. 18

the onset of the "dog days" in August, the element of fire waxes in importance. The characteristic activity of the fire is destruction. (Fig. 18) In the fall, the plants are destroyed and once again consumed by Mother Earth. The fire clears off the summer haze and ushers in the brillance of the fall skies. The earth gives up the heat retained during the summer into the atmosphere, producing Indian summer. Water is drawn out of contact with the earth, reducing its fertility. Crops ripen and are harvested as the earth draws plants back into their roots. Earth energy waxes, fire leaves the earth and cold commences.

The earth body is dense and dark. Her minerals are also dense and resistant to change. Rocks, crystals, bones and teeth all show the dense rectangular forms characteristic of creations developed by strong elemental earth. (Fig. 19)

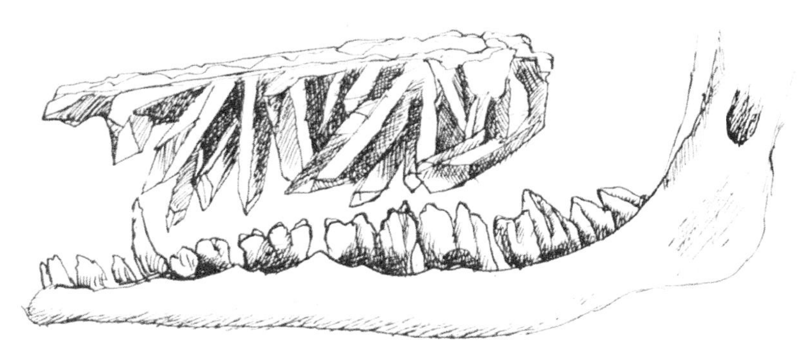

Fig. 19

The denseness of earth opens up when filled with water and dies when sucked dry by air (wind) and heat (sun). As a result, the earth holds great attraction for the water. But its denseness also holds much resistance for the water. This flow and resistance of earth and water give way to a common dance seen in the meandering oxbow creek or the scalloped shore of the bay or river. Sinuous twistings and turnings mark all of the creations that spring from the interaction of the fluid water struggling with its earthy burden. (Fig. 20)

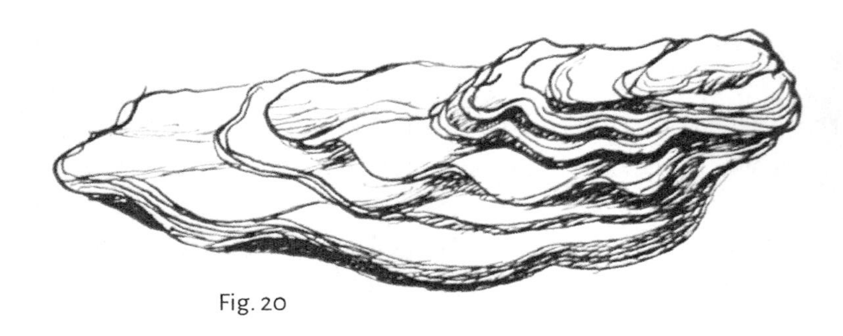

Fig. 20

The veins of your hand and the branches of trees all point to the same source. When the earth is forced to interact with the other elements that it has no affinity for, its own qualities become severely altered and a *Sulf* reaction of oxidation is developed with air; or, if earth is forced to interact with fire, its vital organs (metals) are smelted out. The fire is so intense on the earth that it forms a liquid and tries to escape, thus showing its strong affinity to the water element, the earth's earthly lover. The element of water typifies all that is rhythmic or wavelike in organic nature; its signature is the wave or the half-moon. (Fig. 21)

Water works through gentle and persistent penetration along the path of least resistance. By guiding the interaction of earth and air, water is the great harmonizer stimulating mechanical and chemical reactions within the organic sphere. By expanding and contracting in the presence of heat (sun) or cold (earth), the water

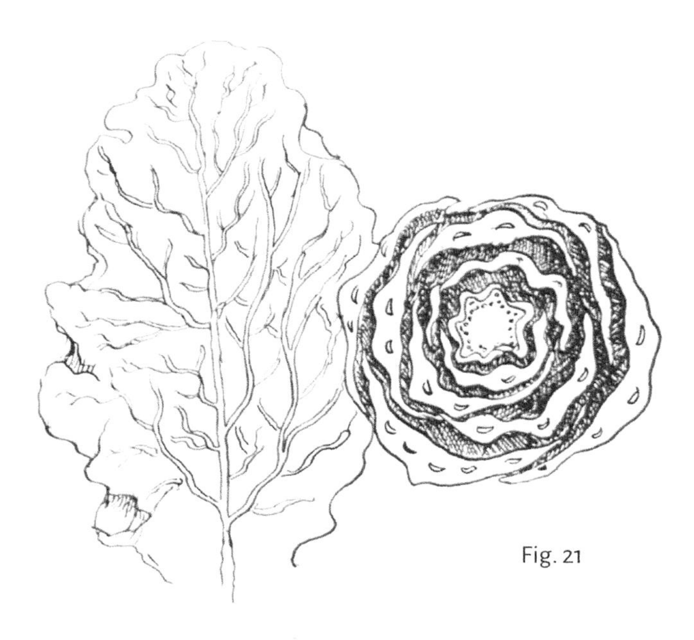

Fig. 21

on the earth is both nature's great builder and her demolition agent. Water gently trickles down into the cracks of a large boulder and at night freezes, cracking the stone into smaller pieces. The rocks are then tumbled down a stream bed until they are pulverized into silt. The silt is then built into plants or settles to form shales and sandstone. In this process the stiff, rectangular crystals are worn down and the dense, intractable earth takes a step toward a higher form through the efforts of its benefactor, the fluid water element.

This cycle gives rise to plant forms by producing stiff, dense, crystal-like roots that exist in a dense, stiff, earth element. The succulent leaf, however, epitomizes the veining, wavy, watery nature of earth moved by water.

The earth and water together constitute the earthly, dark pole.

Here, sightless creatures like worms and slugs crawl on plants without true seeding processes, like ferns and fungi and lichen. (Fig. 22) While earth and water generate organic forms, their heavy, dense, viscous nature must be refined and opened up in order to develop higher life forms. To intensify and refine the earthly elements is the function of the cosmic elements of air and fire. Air/light and warmth form the cosmic light pole, the realm of sublimation and transmutation. (Fig. 23)

Roots sprout leaves, and as leaf follows leaf on the stem, the plant becomes more

Fig. 22

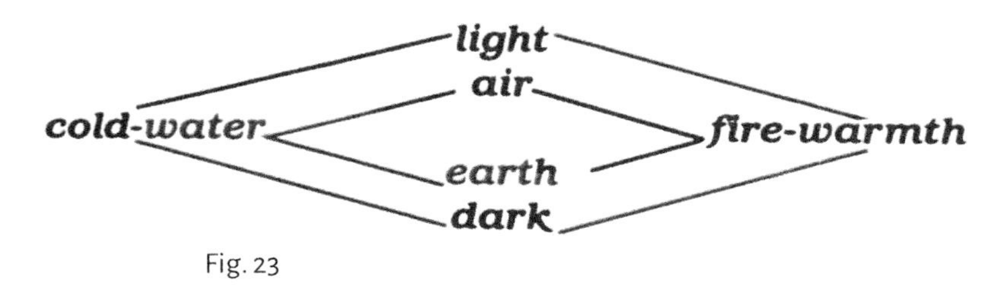

Fig. 23

intensified and refined as it raises its growing center off the ground and into the sphere of light and warmth. The air/light element works to change the gross, dark leaf into the colorful, delicate, translucent flower petal. The stamen, with its thread-like filaments and anthers, also shows the influence of air/light. The air element is strongly ***Mercur***, as is the water element, gas transmission from liquids being the main activity of air and water. The tendency to form membranes is very characteristic of air. The activity of air on liquids in the formation of membranes is easily seen in the formation of soap bubbles. Here, moving air works to bridge the gap as the bubble forms. Balance between inner pressure and outer pressure is maintained and the perfect sphere results. (Fig. 24)

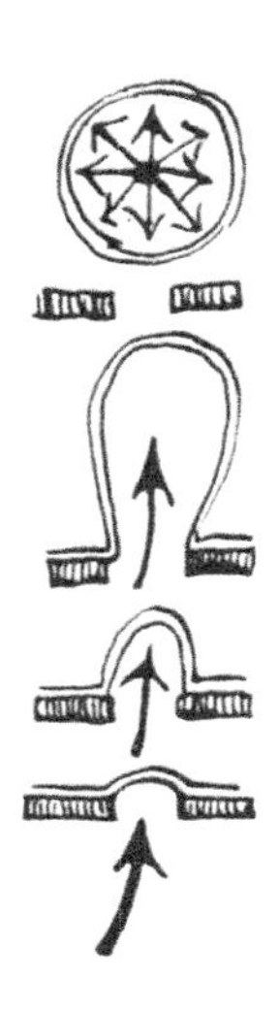

Fig. 24

The addition of alkalized fats (earth) to water adds denseness and enables the air to support the water to form a sphere. Another strong, air-regulated phenomenon is capillary action. (Fig. 25)

As the surface (air) of the water is increased, the air can support and buoy it up until the water becomes so drawn out that it is almost all air (membrane). At this point, the rising character of the air takes over and the water can overcome its affection for earth and for downward motion. Moving air passing across the top of the capillary increases the force of the action. This principle finds its expression in the stems of plants and is most observable during the leafing-out of the trees in spring. Here the tubes in the stem produce strong turgor pressure to swell buds, lifting a tall column of water to great heights. But leafing-out is not a water-regulated phenomenon. It is the wind passing across the ends of the branches that encourages the buds to flourish into leaves. The flat leaf carries the signature of earth and water in its veins, but the light-seeking chlorophyll and the gas exchanging stoma point to the action of the air/light.

Fig. 25

Air and light are not, however, the final stage of elemental transformation. Life needs fire.

Most plants cannot propagate themselves, so they need the touch of a higher life form to quicken their life energy. The heat-loving insects are presented to the plant so that seed will be produced for the continuation of the race. Plants must burn away their lower natures in order to concentrate life essence into the seed. They are guided in this activity by the rhythm of the sun moving before the stars and are serviced in this by the hovering, iridescent, heatloving insect. (Fig. 26) Warmth brings out fragrance and stimulates the seeding process in most plants. The heat swells the bulging fruits and ripens them until they dry out to produce the ash-like seed. The seed can then survive through the cold and damp of winter because of the contained warmth.

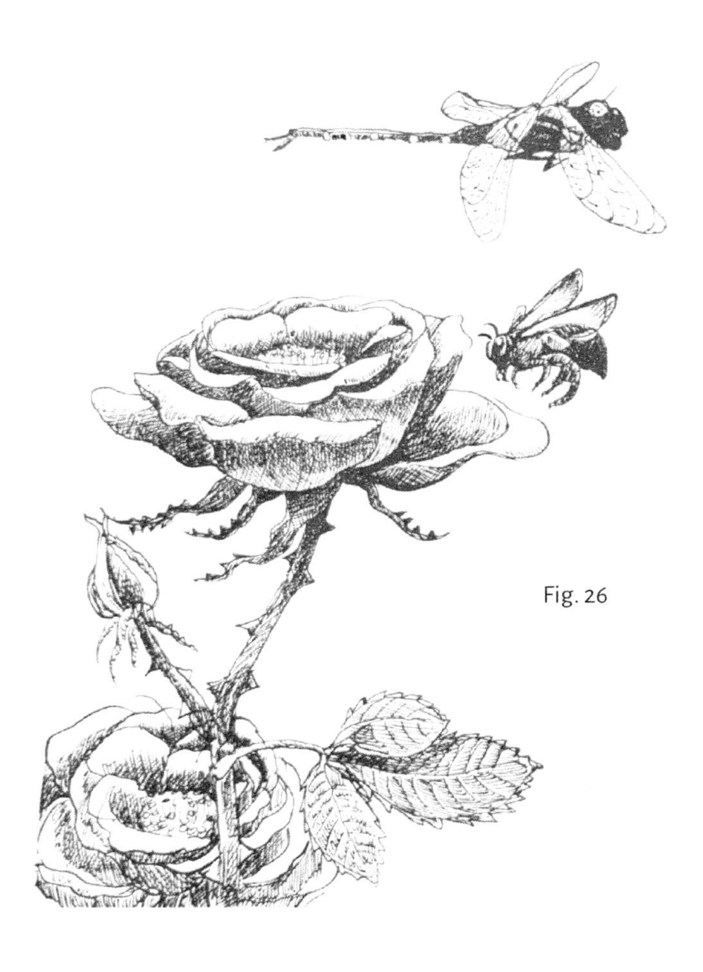

Fig. 26

Planters know that damp is the worst enemy of seed viability. This is because the fire in the seed finds it enemy in water.

In plants where seeds are produced, heat develops oils through sublimation. Elementally this is an example of the fire cycle, which is a *Sulf* reaction of destruction. In nature, the order of the four elements is consistent with earth being at the base while water is in close connection slightly above earth. Next comes the air, sharing a common sphere of activity with the water and transferring sensitive impulses between elements, and finally the warmth or fire, hovering above all the others, quick to escape upwards in an effort to return to the sun.

Fig. 27

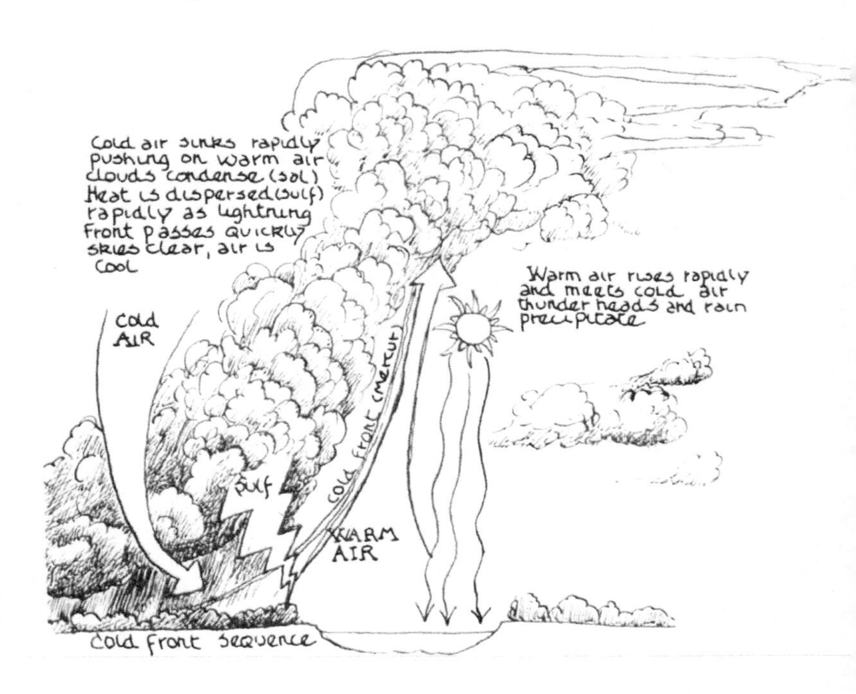

Whether we observe puddles, ponds, lakes or oceans, this order is at all times the goal of the etheric formative forces. Plant life, animal life and weather phenomena have their beginnings in these relationships. Earth holds and contains water that is acted on by the warmth of the sun.

The water and air interact in a ***Mercur*** reaction and the water rises. (Fig. 27) In order to stream back to the sun, the water must leave its coldness with the earth and take on heat. Rising with the heat, the water meets a layer of cooler air and begins to revert back into a more earthly form; it seeks to retain the wholeness it possessed when in contact with the stable earth. (Fig. 28) Clouds are brought into being (*Sal*) through this yearning, and the water condenses and gains mass.

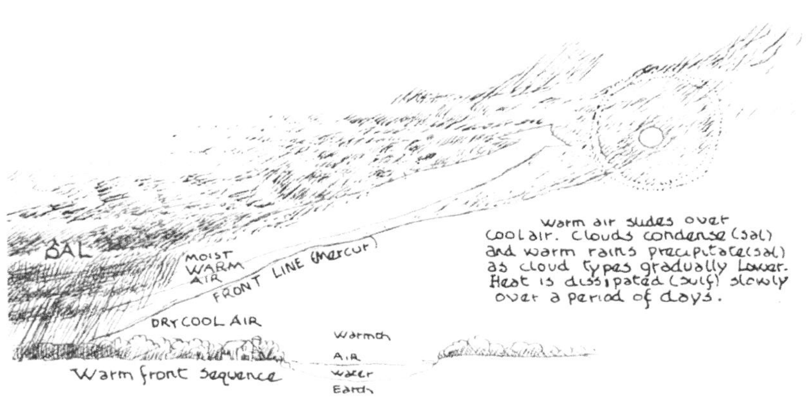

Fig. 28

This water is in an environment that is alien to its elemental nature. It yearns to touch the earth and make it live. The water already present on the earth sets up a strong attraction to water in the air and so it rains. The heat stored in the air/water is released and dissipated (*Sulf*) as lightning. The heat of this reaction triggers an intense precipitation of water from hydrogen and oxygen, thus producing the cloudburst. Nitrogen is also precipitated into the falling rain which soon feels the embrace of the cool, dark earth. The element of fire is present as the energy required to lift the earth/water and is dissipated in a strong *Sulf* reaction in the thunder and lightning. These cycles of cold, warm, wet and dry rotate endlessly in the atmosphere.

In organic chemistry, the concepts of warm, dry, wet and cold can be used to describe the relationships among the various elemental substances in nature's household. Within the element earth there are various "earths" that have relationships with water, air and fire. (Fig. 29) For instance, calcium and magnesium

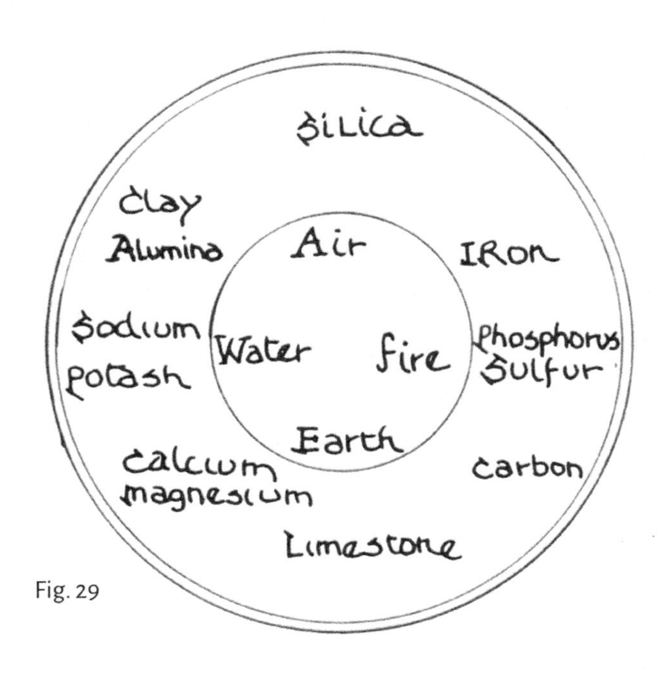

Fig. 29

are earth alkalis. In nature, they occur together in compounds and support each other in their insolubility in water. Excessive fertilization with calcium has a tendency to inhibit the activity of magnesium. Their relationship is reciprocal, yet both want to retain their dense earthiness at all costs.

Potassium and sodium, on the other hand, are mineral alkalis that possess strong attraction to water. These two salts govern the flow of water and other fluids in the body.

Potassium, which is found inside the cell fluids, is the sensitive salt always seeking to support and balance plant growth in an orderly fashion. Found primarily in roots and stems, it works to give expression to the limestone. Sodium, which is found in the fluids outside the cell, is bold and assertive in strong *Sal* reactions. (Fig. 30) Found primarily in the fluids in the animals on earth, sodium is heavy and forms a large part in the clay of the earth. Sodium is more watery, dissolving and flowing into solution, and potassium takes a more balanced role in osmosis.

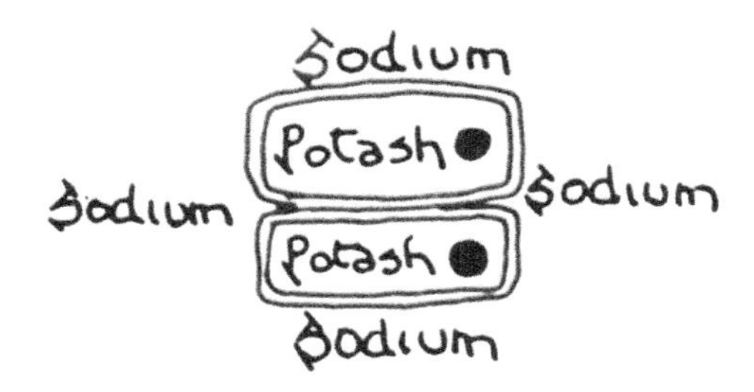

Fig. 30

Potassium relates well to the influences of calcium carbonate or limestone. Sodium on the other hand relates well to the silica pole, to form silicates of soda (water glass).

Alumina (a heat resistant, neutralizing metal) and silica form $AL_2O_4\text{-}SiO_2 + 2H_2O$ (Kaolin). The water alkalinity of sodium and the airy crystalline density of silica meet and merge in a *Mercur* material capable of supporting great change. Although as a mineral silica is dense and insoluble, molecularly it combines well with other minerals and gases, especially when influenced by its cosmic partner, fire. In order to facilitate the union of silica and sulfur, iron acts as a flux to lower the high melting point of silica. (Fig. 31)

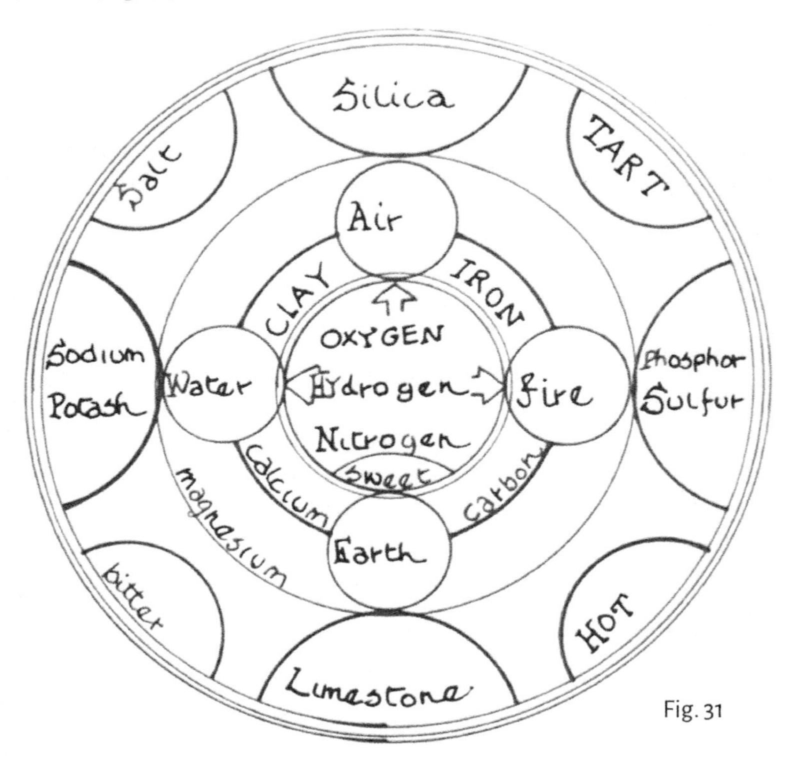

Fig. 31

Both silica and iron have strong affinity for oxygen, both are eager to become acidic oxides by attracting and using oxygen. The "light-bearers", as Rudolf Steiner called phosphorus and sulfur, consume the completed form of life, giving off ash (carbon), or earth, gases (O,N,H), water/air, and energy in the processes of destruction, sublimation (fire).

This burning of the earth results in carbon. By its sheer emptiness, carbon is a mineral capable of great possibilities for organic development. Carbon is a study in extremes, dense and dark yet extremely porous and featherlight. Apparently lifeless, yet capable of infinite chemical bonding, carbon is the true keystone in the forms of living beings. As the fiery complement to the plastic clays, carbon is the great plastic artist of nature. Strongly *Sal* (absorbing, precipitating), carbon builds the bridge between the *Mercur* nitrogen, oxygen, and hydrogen to form the carbohydrates, sugars, alcohols and fats. When carbon meets a solid *Sal* substance in calcium, instead of the *Mercur* gases, its own *Sal/Mercur* forces are blended with those of calcium to produce the limestone craving in the soil and so the cycle begins again. These elemental relationships are repeated endlessly in all the natural phenomena of life, death and rebirth. Nowhere is the elemental rotation more evident than in the life cycle of a composting heap of refuse. Here earth, water, air and fire each play an integral part in the breakdown of raw organic substance into a stable humus capable of promoting life and health in plants.

Mercur compounds (catalysts)

carbon + hydrogen = methane
nitrogen + hydrogen = ammonia
carbon, hydrogen, oxygen = sugars, starch, oils + fats

carbon, hydrogen, oxygen, plus nitrogen = proteins
silica + alumina = clay (neutralizer)

Sal compounds (precipitators) alkalines

carbon + calcium = limestone
carbon + calcium + phosphorus = bone meal, oyster shell
carbon + calcium + potassium = wood ash
sodium + nitrogen = chemical fertilizer (salt petre)
sodium + chlorine = table salt

Sulf compounds (burning, oxidizing, acids)

carbon + iron + phosphor + oxygen = steel (slag)
Oxygen produces acids through oxidation.
Iron attracts oxygen to aid oxidation.
Proteins (C, O, H, N) + phosphorus = energy
Oxygen gives life to Air.
Nitrogen gives life to Earth.
Hydrogen follows water and reacts violently to fire.
Phosphorus and Sulfur are involved in protein assimilation and destruction.
They are found in animal bodies and manures.

Most manures, garbage, plant remains, animal remains and crop residues are considered elementally as earth. The living entities that have formed the earth into their bodies have left the earth locked up in the bodies that they discard. Living bodies have formed the earth into unstable compounds, however. These proteins have a tendency to become extremely *Mercur* in order to allow the organism to lead a sense-oriented life and to have freedom of movement. Upon death, the manures or residues begin to react in a strong ***Sulf*** reaction of rotting. The ***Mercur*** compound of ammonia is freed by this *Sulf* reaction. Ammonia carries off with it a great amount of the nitrogen in the rotting refuse. This loss of the ***Mercur*** nitrogen can be prevented by the careful manipulation of the heap in order to balance it elementally.

Fig. 32

Rock powders added to a heap are a simple means of directing nitrogen into stable humus. Granite dust and limestone help heaps made with strong phosphorus materials like chicken manure or feathers or fruit pomace. These manures are hot, or ***Sulf*** earths. (Fig. 32) They need sweetening. The cold, or potash, earths are pig, cow and sheep manure. These earths need a ***Sulf*** reaction in order to be balanced. Phosphate rock or the droppings of fowl add some heat to the cold earths. Horse and cow are the most elementally balanced manures, so they will compost by themselves into a fine product. Many "earths" like spoiled vegetables, kitchen garbage and urine-soaked bedding add a substantial amount of water to the heap. Grass clippings, on the other hand, add almost no water, making additions of water necessary. These contain strong earth (silica) and very little water. The ideal amount of water is about the amount of water in a fresh cow pie. Too much water and the bacteria doing the initial breakdown start to suffocate. The heap becomes sodden and foul smells escape, giving indication that a rampant *Sulf* process is burning off the nitrogen and dissipating the life forces in the heap. The surest remedy is to push ventilation shafts into the heap with a pole. With a good supply of air and water, the bacteria proliferate and their metabolic energy begins to heat the pile. The heating can become too vigorous, especially if hot earths and highly nitrogenous substances like chicken manure (N 20/P 16/K 8) or blood meal (N 15/P 1/K 1) are used. A gentle fire stage is the ideal, so that the fire does not burn the fertility out of the compost. If, on the other hand, wet and cold earths predominate in the heap's make-up and the weather turns wet and cold, the heap could easily become soggy and smells could develop. Quick or burned lime can then be sprinkled down into the ventilation shafts. The intense fire of the burned lime offsets the soggy denseness of the cold earths, and balance is restored as the smells go back into the heap, taking the *Mercur* nitrogen with them.

Chapter IV: Plant Life on the Earth Body

The body of the earth constitutes the earthly pole. The starry heavens and the solar system form the opposite, cosmic pole. Among minerals, limestone and potassium are earthly, silica and phosphorus are cosmic. These are the main agricultural minerals. The alkaline earth, limestone, has the strong ***Sal*** (absorbent power) of calcium and carbon. These forces serve to draw nutrients, gases and water into the soil This absorptive (***Sal***) quality is an aid to legumes in the task of fixing nitrogen in the soil.

Looking at the pea, we see a picture of the typical lime-lover. The limestone draws at the roots, causing them to start to branch out more and more finely. The fruiting process in turn moves down into a lower area and swells the broad stem; fleshy, hairless leaves clasp the swollen stem while the fruiting process is drawn still further into the roots where we can see nodes attached to the roots like little peas. The legume is a blue-print for the limestone process in soils. (Fig. 33)

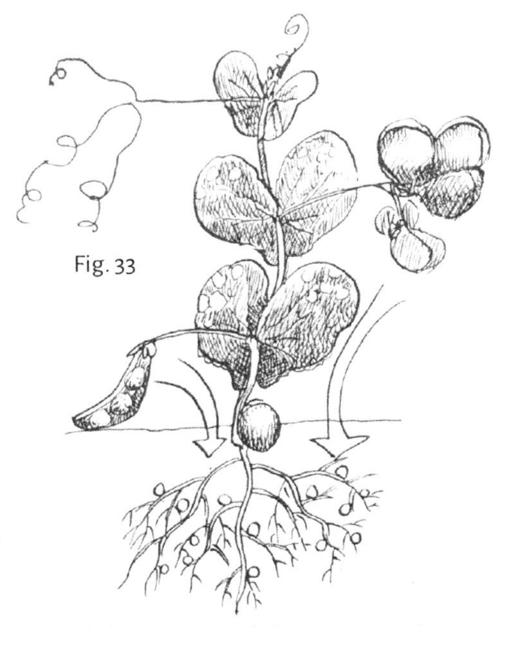
Fig. 33

Potassium is another *Sal* alkaline. Due to its position deep within cells, potassium typifies the precipitating processes of the *Sal* minerals. Swelling stems, ribs, or nets typify potassium. Potash and time are often found together in nature in feldspars and calcium compounds. Garden plants which need lime/potash also bear these ribs or swelling stems. These plants include leeks, onions, beets, cabbage, chard, parsnip, celery, celeriac, and stemmy herbs like sage and summer savory. Lime applied lightly before sowing is a great aid to these plants. A list of definite lime/potash-lovers could include asparagus, cantaloupe, cauliflower, cucumber, iris, fennel, lettuce, parsnip, rhubarb, salsify, squash. Flowers that prefer lime are sage, aster, penny cress, bluets, chamomile, yarrow, tansy, lily and sweet pea. In general any plant which has a tendency to develop strong stems with fleshly, clasping leaves, hairless surfaces and a fibrous, diffuse root mass, could tolerate a sweet soil. Potash is very important to the fruiting process and is most critical in the formation of the root system, the formation of woody trunks and any structure where rigidity is needed to support soft tissues. Stems swelling for storage need potassium. (Fig. 34)

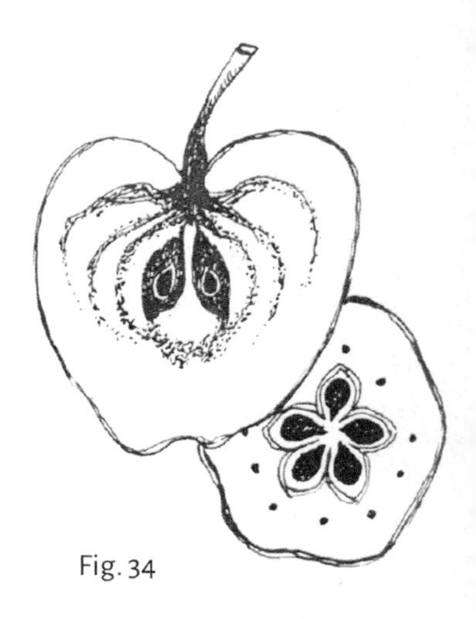

Fig. 34

Sources of potassium and lime are most abundant in the form of ground rock powders. Ground granite, marble chips, mica, marl, basalt and wood ash are the most common. Bone meal contains potash balanced with phosphorus and calcium. Wood ash contains 7% potash and 30% lime. Potash feldspars contain large amounts of potash and very little lime. Greensand and marble granite dust and basalt flour are slow release forms of potash.

Lime or wood ash mixed with manure will work to liberate the nitrogen in the manure. For this reason it is best to add potash to manure just as it enters a compost or just as it is turned into the ground. This fixes the nitrogen into the humus of the compost with minimal loss into the air. Plants are aided in potassium absorption by the sprinkling of an herbal tea made from the yarrow plant.

Quicklime is a caustic form of limestone made by burning lime in a kiln. The resulting product is the basis of common whitewash and will control odors and sourness in a compost if used sparely. Do not use quicklime on plants or soil, however. (Fig. 35)

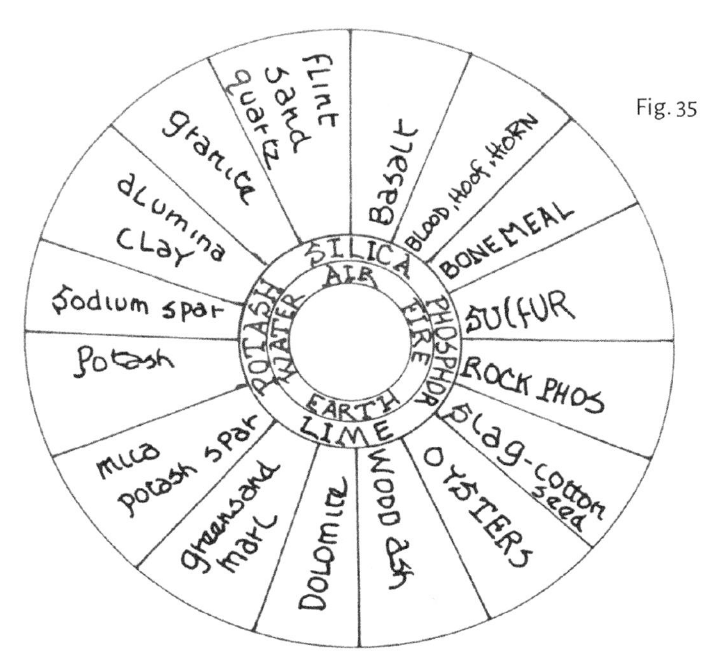

Fig. 35

At the opposite pole in the earth body are the cosmic minerals, silica and phosphorus. Silica is found in nature as sand, flint or quartz, or as silicates in feldspars and mica; as such it is the most abundant mineral in the earth body. Phosphorus is the hottest mineral in the earth. Together they are the complement to the spreading, watery energies of potash and lime. Silica is related to elemental fire. They thus produce forms that are light, hot, thin and brittle. Plants rich in silica radiations have long, thin, brittle stems, finely divided foliage, and fine sharp hairs on leaves or stem. These plants are the acid-lovers, like the pines and grasses. Flowering evergreens, like rhododendron, azalea, and holly, show the brittle stem and flowering process indicative of the cosmic silica/phosphorus processes.

These forces, when strong in roots, produce a stringy rhizome instead of a ramified root. Characteristic silica/ phosphorus plants would be the grasses: long thin stems rising from a rhizome, pointed silicious leaves rising quickly above the earth. (Fig. 36)

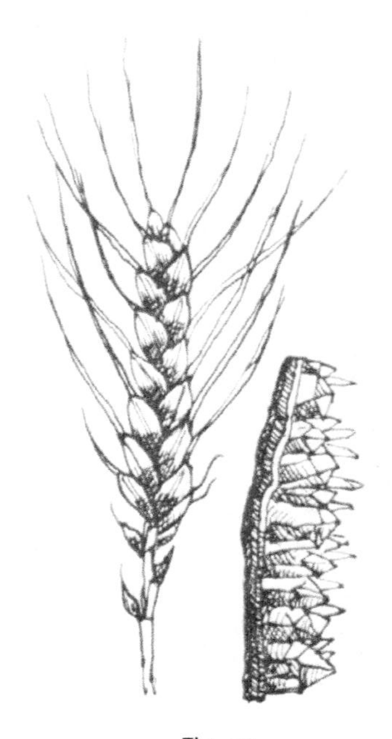

Fig. 36

In grasses we see the picture of a plant trying to rise up and away from the earth body. This rising is so strong that the grain forms prop roots showing the presence of root forces up on the stem. This is the opposite tendency from that of the legume, which pulls the fruit process down into the root zone. The action of silica works through air/light to refine and enhance the grossness of leaf and root into the fineness of flower. The silica forces develop the plant in the vertical, intensifying stream, rather than in the horizontal swelling of stems or thickening of the leaves (as with potash/limestone). Roots tend to be compact, woody and entire rather than diffuse, branching and tender. The tendency for stem/root processes to be intensified higher up in the plant is so pronounced that grasses like quack grass and knotweed have forsaken seed viability in favor of strong reproductive power in the rhizome. In other words, the seed force is completed in the root/stem area and has no need to rise into the inflorescence; thus the root/stem becomes the seed, and the flower seed is sterile. The power to do this comes from the intense formative power in the hexagonal quartz crystal. These crystals give plants the reflected forces of the cosmos. Whereas limestone is typically organic in origin, with strong lunar forces of earth and water, the silica/phosphorus is filled with the light/ warmth of the sun, and these elements transmit the light and the heat to plants for flowering/seeding formations. Plants then are strung between the forces of the lime/ potash, sucking downward, and the silica/phosphorus forces pushing upward from below. Plants which favor silica and the acid pole are being pushed upwards from the dark, damp soil into the light and warmth where their lower organs can be metamorphosed into higher forms capable of reproduction.

By observing a group of root vegetables we can get an image of the subtle influence and interaction of the two opposing tendencies of silica/phosphorus and limestone/ potash. (Fig. 37)

In the radish, the most acid root, the phosphorus process develops hairiness in the leaf, stringy, tough roots, and a tart, hot flavor.

Fig. 37

The carrot demands pH to be balanced, but its crystalline root and finely divided foliage show silica relationships. The turnip has the stringy taproot of the acid lover, but the balanced, swelling shape of the bulb is a ***Sal*** indicator of potassium. The beet has a stringy root, but the swollen stem and strongly veined succulent leaf show a strong limestone tendency. The parsnip with its succulent, ribbed, celery-like stalk and sweet, fleshy root shows strong affinity to the limestone forces. ***Sal*** forces are revealed in the bulb shapes of the Allium family. (Fig. 38) Here the bulb is pointing up as if the energy were being pulled down by the lime into the swelling stem. In acid roots, like the radish, the bulb points down as the plant's energy is directed away from the surface of the earth by the silica.

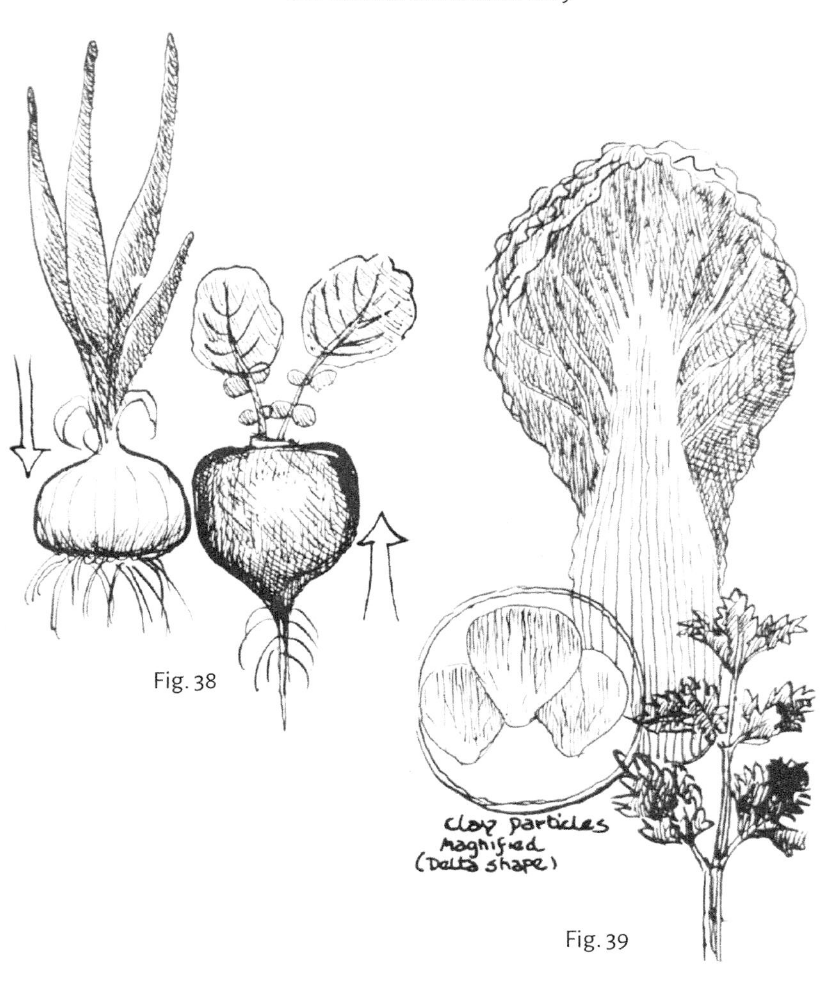

Fig. 38

Fig. 39

Among leaf vegetables, those leaves having smooth margins and wrinkles show limestone energy at work, Hairy leaves like the nightshades, nettles and grasses show the influence of acidic silica. In general the shapes of the leaves will be needle-like for acid-lovers and smooth and wrinkled for alkaline plants. Plants with large, water-filled leaves are exhibiting a gesture associated with the third most abundant mineral condition in the earth, the clay. (Fig. 39)

Acting as the mediator between silica and lime, clay impresses the leaf with the delta -shaped shingles of its own particles. Where we find needles, we find phosphorus/silica. Where we find veins, there is potassium/limestone. Where we find fleshiness, there is nitrogen/clay.

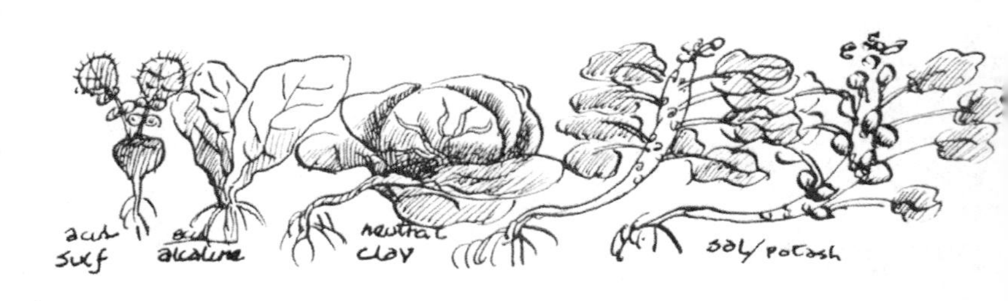

Fig. 40

Within a single family of plants (Cruciferae), these characteristics may be observed. (Fig. 40) The ***Sulf*** crucifer is the acid loving radish. In the radish the root forms quickly and before we can blink an eye the radish is flowering. This shows that the influence of silica/phosphorus is predominant. At the opposite pole to the radish we find the stem/leaf crucifer, the kale. The kale develops a strong potassium stem gesture even in the leaf. It is a plant concerned with becoming instead of flowering and so it flowers only in its second year. The cabbage, another cousin of the radish and the kale, retards flowering in a strong ***Sal*** gesture but shows only a passing tendency to develop strong stem forces. The cabbage is the root/leaf crucifer and its life gesture is to grow as rapidly as possible, producing leaf after leaf. Thus the ***Mercur*** forces of nitrogen and clay are especially dear to the cabbage. Potash is present in the stem and strong ribs of the leaves, limestone in the fibrous diffuse root, and phosphorus in the hot taste. From this we can see that the cabbage needs balance or else its ***Mercur*** tendencies can easily lead to rot or stunted growth. From this we can see that crucifers have developed distinct body gestures valuable to gardeners. The radish is a root/flower crucifer with a strong ***Sulf*** (burning) process. The mustards have alkaline roots and acid leaves and are quick to bolt to seed like the radish. The cabbage is the balanced root/leaf crucifer. The kale and collards are stem/leaf crucifers and appreciate a bit of extra potash. Brussel sprouts and broccoli are stem/flower crucifers and phosphorus balanced with potash hastens the development of the flowering process.

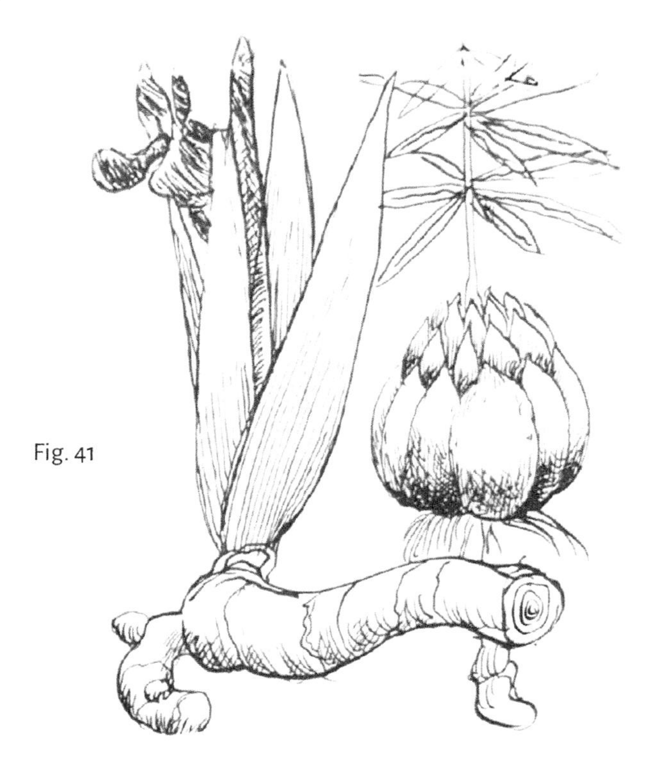
Fig. 41

Similar divisions can be observed among flowering bulbs. In the phosphorus/silica, acid-loving lily bulb, we find crystalline, finely divided scales, thin leaves and narrow stems. (Fig. 41) In the sweet-loving iris, we find a succulent rhizome, concentric layers, broad leaves, and strong stems.

The acid-lovers, like marigold, hawkweed, knapweed and flax show thin stems, feathery leaves, stringy roots and profuse flowers. The alkaline flowers exhibit tendencies to fleshy leaves, like the cresses and sage; broad stems, like mint and summer savory; and blue flowers, like iris, bluets, phlox, or red-blues like carnation and sweet pea. Trees showing acid preference are the evergreens like pine and fir, with their finely divided foliage, scaly bark, thin trunks, and stringy roots. Extreme acid-lovers like rhododendron and azalea (Fig. 42) show extremely thin, burnt and woody stems, scaly bark, and stringy, tough roots. Alkaline trees like the beech show smooth, broad trunks, as do maples and many fruit trees.

Fig. 42

In the fruiting plants we can see the marks of the two polarities most clearly. (Fig. 43) Acid fruits include most berries like raspberry, blackberry, and strawberry, each bearing the characteristic hairy, thin stem; in the case of the rose, another fruit-bearing acid-lover, the hairs become thorns. Other acid fruits are the blueberry and huckleberry, which have the thin, knobby-ended stem of the rhododendron, and a tart, acid flavor.

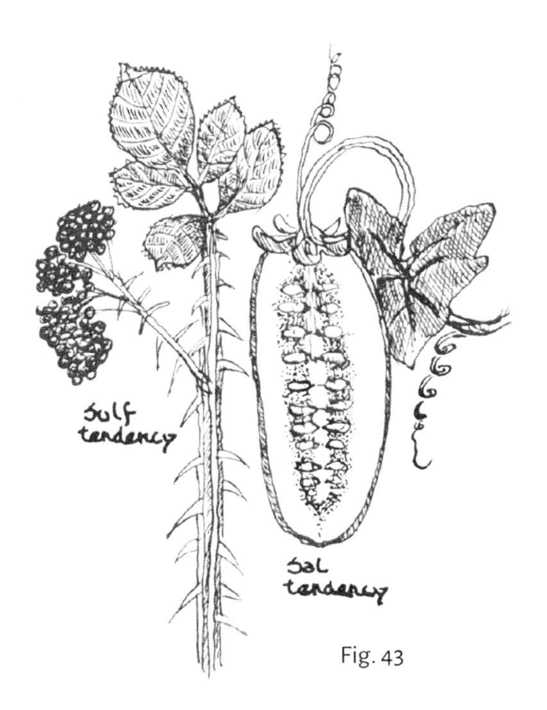

Fig. 43

Sweet-loving alkaline plants among the fruits are generally those exhibiting veins spread out in nets through swollen or mealy fruit. These include squash, cucumber, plum, cantaloupe and apple (an alkaline Rosacae).

The veining tendency arises in the limestone/potashalkaline polarity. It is found as a dominant gesture in the leafy crops like spinach, chard, rhubarb, celery, celeriac, parsnip, beet, lettuce, cauliflower and cabbage. (Fig. 44)

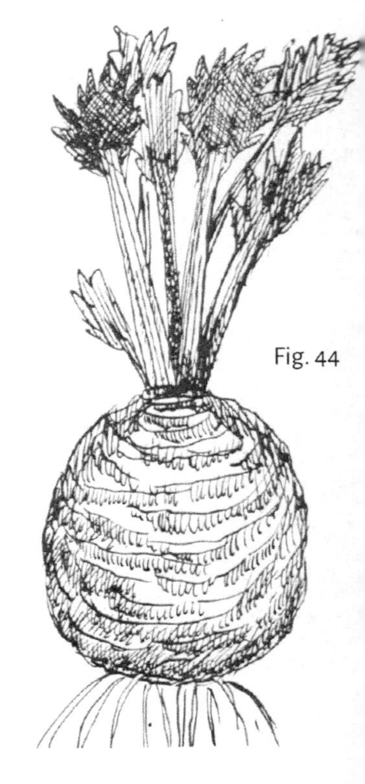

Fig. 44

Ribs, ridges, nets and veins -especially near the stem region-tell of strong lime/potash activity in the plant. Potassium also works to develop swelling roots and stems, and swellings on plants in general, through strong precipitation (***Sal***) tendencies.

The idea of transmutation must be kept in mind when observing plants like squash and cantaloupe. These fruits require strong potassium for large fruits and long vines, but strong potassium formations like these must also be balanced by strong phosphorus/silica radiations if flowering is to develop and fruits are to ripen. Thus the calcium-rich leaves of the melon are covered over with perfect, sharp, silica needles. Plants like the pine have characteristics that show the strong influence of silica in needles, pungent sap, stringy root and scaly bark, balanced by an all-pervading spiral motif that is strongly characteristic of limestone productions. In the pine, lime and silica blend in balance of form. Any attempt at a holistic study of these influences on plants is best undertaken by keeping balance and transmutation firmly in view.

It is of importance to note that these tendencies by no means form a clear-cut division in the plant world. The plants live suspended between the two poles, trying to balance the extremes in order to survive. Nature has provided a number of processes to enable the plants to do a this. The first is the intensification process. Silica shows strong relationship to the phosphorus pole. But silica is not found evenly distributed throughout the plant. It is usually absent in the root/leaf formation and found in greater and greater abundance as the plant approaches flowering. The silica is intensified as the energy of the plant becomes intensified for flowering. This process can be observed in a single leaf of an acid-lover like the radish (Fig. 45): fine silica needles emerge intensely at the upper tips of leaf lobes. In acid vines like potato, tomato and watermelon, the needles are dramatically intensified as one approaches the blossom. In like manner, potassium becomes intensified by the flowering parsnip into ridges that turn into the rays of an umbel flower and calcium is intensified by cucurbits. In these plants, the minerals become intensified through the growth process. This intensification sets the stage for the mysterious process of transmutation. This occurs in plants like the oak and the horsetail. Both of these plants live in barren, sandy, acid soil, and yet they both contain the richest store of organic calcium in the plant world.

Fig. 45

Where does the lime come from? It is produced by the process of silica intensification. The horsetail (Fig. 46) has 98% silica in its ash. This super-intensified silica cannot become more acid and so, in one of nature's great feats of balance, the extreme acid condition is transmuted into an alkaline condition. In these plants we can see a model of all processes of transmutation of substance. Opposites must balance, or else life is extinguished. Balance is maintained at any cost, as extremes intensify into critical conditions that must foster a complete transmutation.

Fig. 46

It is here that the clays have an important function. Clay possesses alumina that seeks to balance extremes of acid and base. This neutral function is also a position. Clay helps mediate between silica and limestone; it possesses tremendous water-absorbing qualities and water-distributing potential. Wherever it is found in nature, clay aids in the retention of water in earth. Clay underlies many stream beds, absorbing water and releasing it slowly, distributing it throughout the soil layers. Clay gives plasticity and colloids to the rigid silica formations. Clay also gives adhesion to the dissolving chalky limestone, thus tempering the alkaline "hunger" in the soil.

In the plant world, clay gives its shingle-shaped particle form to the leaf formation. The clay's activity as a mediator is passed on to the leaf as mediator between the poles of root and flower. Plants known for their leaf development have particular affinities for clay. These plants include cabbage, chinese cabbage, cauliflower, kale, onion, leek, lettuce, parsley. Just as clay guides water through the earth, the leaves guide water through the plant.

Keeping this in mind, it is possible to construct various "biotic" substances or teas that can be given to plants with the intention of controlling the development of any specific organs or life gestures which we wish to cultivate. A few simple rules must be adhered to, if predictable results are to be maintained.

The rule of conservation of life energy is most basic. In essence, this rule states that the energy forms assumed by a living being are found in potential in its remains. For example, Wolf Storl, in ***Culture and Horticulture***, states that manures from animals will most easily grow the plants that the animal was fed. Manure from hens which are seedeaters tends to be high in phosphorus and is therefore perfect for seed-bearing plants. Cows, horses, goats, geese and sheep eat grass. This makes their manure excellent for leaf crops and pastures but less suitable for cereals. Pigs and cattle fed on roots produce a potassium-rich manure perfect for the fertilization of crops like onions, leeks, celery and parsnips. The manure then has a "specificity" in regards to plant development. Plants grown with such manures grow like the food of the animal producing it. The energy is transmitted in "blocks" from soil to plant to animal and back to plant *via* the soil. In this system, entropy is minimal. Forms merely undergo transmutation with the patterns extending beyond life and death. The plants take in the earthly and cosmic elemental forces and raise them to a higher vibration of life. The animals then eat the plant and act as transformers to step up the current to an astral level. (Fig. 47) This higher vibration is passed through the manure to the soil and then back to the plants. If the manure is not fixed into a compost, this high vibrational energy often becomes dissipated and lost. Hence the dynamic herbs come to our aid.

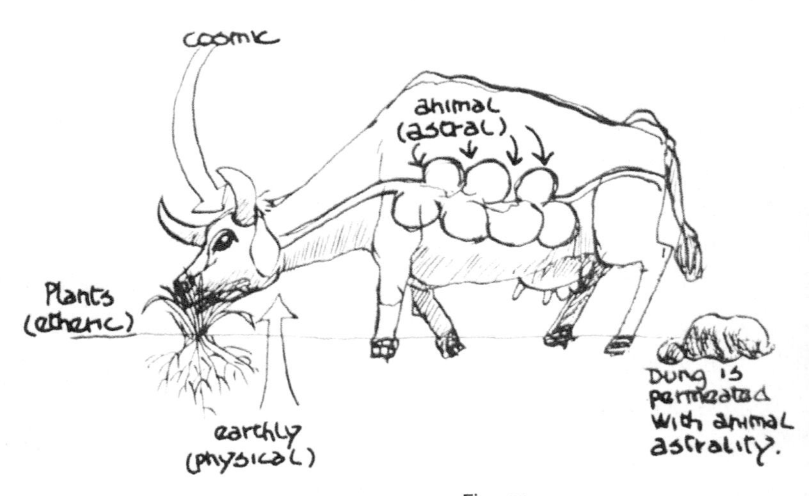

Fig. 47

Herbs are plants which have arranged life energy into very specific and highly efficient patterns. These patterns or gestures can be used to guide the firing of life energy in manure into a stable form suitable for use as fertilizer. Fermentation of the manure in a barrel of rainwater can be custom-controlled for specific plants. For instance if lettuce, cabbage, and chard (all calcium-lovers) need fertilizer, you can steep a gallon of manure from cows, sheep or horses in five gallons of water. To guide the reaction, finely chop two pounds of nettle leaves, two pounds of violet leaves or kelp and two pounds of cabbage or buckwheat leaves, and ferment. These more dynamic herb leaves will swell the leaves of the cultivated crop. When beets, broccoli, kale, cauliflower, collards, rape, mustard or chinese cabbage are desired, fowl droppings and horse or cow manure may be fermented with violet leaves, shepherd's purse, pigweed, and kelp to develop phosphorus and potassium in just the right relationship for stem/ flower development. (Fig. 48) Looking at shepherd's purse, one can almost picture mustard, cabbage or kale. Fruit crops like squash and pumpkin have great phosphorus requirements for ripening fruit and great calcium/ potassium requirements for swelling fruits and leaves. Herb plants that have this life gesture of a rigid, hairy stem, with copious fruiting, flowering, and seeding are the phosphorus-gatherers like comfrey leaf, Jimson weed, valerian and the seed heads of nettles. Fowl droppings rich in phosphorus are fermented with these calcium/phosphorus herbs.

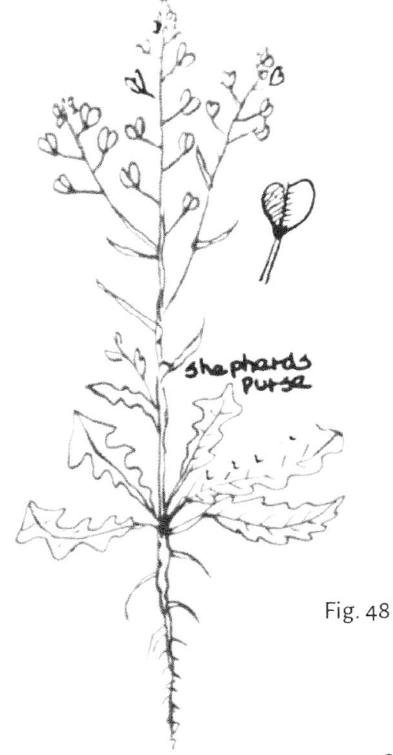

Fig. 48

A revolutionary way of using an understanding of the Earth Logos for fertilization is outlined in Rudolf Steiner's agriculture course. By taking yarrow, chamomile, nettle, dandelion, valerian and oak bark, and subjecting them to "potentizing" forces within the earth, the herbs become highly charged with their own power, in some cases intensified by particular animal organs or parts. (Fig. 49) The resulting potentized substances are inserted into the compost heap in order to enhance and intensify the activity of the heap. The bio-dynamic movement now makes these "weprations" (502-507) available commercially, along with field sprays ("horn-manure" and "horn-silica", 500, 501) and fruit tree sprays.

In general, herbs used for ferment teas must be dynamic in character and can be relied upon to give the plant they are fertilizing strong gestures similar to their own.

Below is a list of the more common herbal regulators and their use.

Alfalfa/clover - high in potash, nitrogen and phosphorus; good for potatoes, corn and seed crops.

Buckwheat - high in calcium and phosphorus; good for stem/ leaf/seed.

Bracken fern - high in potash; strong *Sal* forces promotes absorption of water, aiding germination.

Comfrey leaf - high in phosphorus and calcium; good for fruiting plants-tomato, squash, pepper- and seeds like corn and other grains.

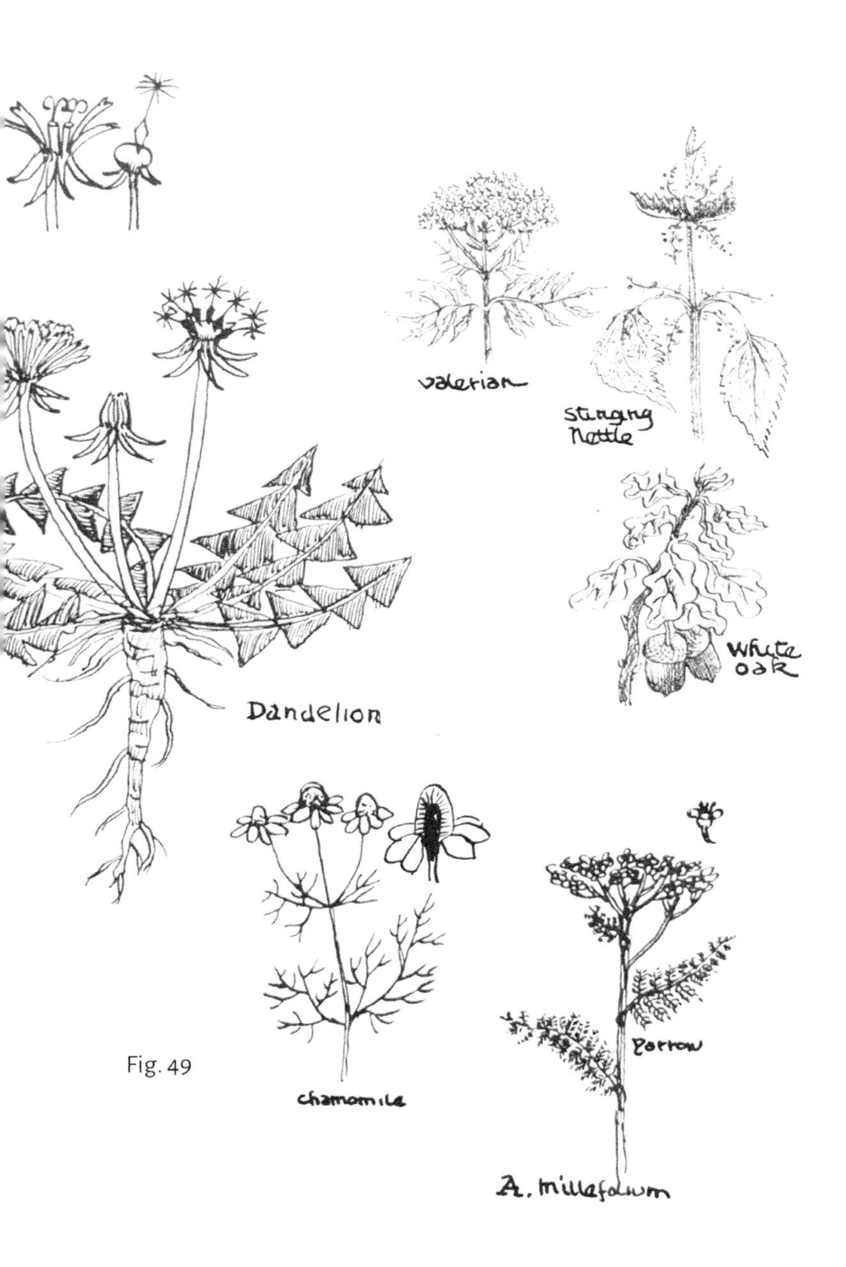
valerian
stinging nettle
White oak
Dandelion
chamomile
yarrow
A. millefolium

Fig. 49

Dandelion - high in silica, potash in roots and phosphorus in flower and seed; good for leaf and root crops, especially biennials.

Horsetail (Equisetum) - strong calcium and silica; good for fruits and vines to combat fungus.

Chamomile - high in calcium, potash and sulfur; good for leaf crops and flowers; promotes health of vegetables in general. Stinging nettle-high in phosphorus and iron; guides organic breakdown in fermentation.

Shepherd's purse - potash root, high sodium and phosphorus and sulfur; good for broccoli, kale and stem/ flower plants, mustards and cabbage if used with potash and adequate nitrogen for heavy feeding.

Pigweed, violet, mallow, lamb's quarters - high in potassium and calcium.

Yarrow - sulfur and potash. Aids potash uptake.

The goal of nature is to give to the earth a balanced vegetation with a wide range of plant types, each suited to a particular set of variables. This insures that the earth will be covered, no matter what the soil condition. When a gardener enters into such a balanced system, it is of the utmost importance that he try to work along the lines that are already established in the natural order. Too often what a gardener puts into the soil upsets this order, and insects and rotting produce are the result. Plants vary in their ability to withstand adverse conditions, so a working knowledge of the various soil amendments and their properties is very useful. Soil additives that increase acidity are peat moss, wood chips, sawdust, sphagnum moss, oak mold, pine needles, pulps, fermentation residues, poorly aired compost, and half-rotted green manures, especially clovers and alfalfa. Rock phosphate and bone meal are also acid minerals.

At the other pole, soil amendments that provide an alkaline environment are crushed limestone, wood ashes (30% lime - 7% potash), greensand (4% lime - 7% potash - 20% iron - 50% silica and clay), basalt, granite, mica and dolomite. Limestone is a rock that has origins in organic life, therefore the soil animals, worms and insects contribute the largest lime deposits in the garden.

To a basic tea made of kelp, nettle and mixed manures the following herbs may be added to develop vegetables in a dynamic way.

Beet - lamb's quarter, shepherd's purse (one part woodash, one part phosphate).

Cabbage - chamomile, yarrow, buckwheat, violet (one part granite dust, one part phosphate, one part clay).

Carrot - yarrow (one part greensand, one part granite dust, one part phosphate).

Corn - clover, alfalfa, chamomile, bracken (two parts limestone, two parts woodash, two parts cottonseed meal). Chicken manure beneficial because of phosphate.

Cucumber - buckwheat, comfrey, chamomile (two parts woodash, one part rock phosphate, one part bone meal).

Kale - pigweed, lamb's quarter, mallow (two parts woodash, one part phosphate).

Leek - chamomile, yarrow, buckwheat (two parts woodash, two parts phosphate, one part bone meal).

Lettuce - chamomile, buckwheat, violet (two parts woodash, one part rock phosphate, one part bone meal).

Parsley - bracken (one part woodash, one part clay).

Parsnip - burdock (one part woodash).

Pea - clover, alfalfa (one part woodash). Inoculate your seed.

Pepper, Tomato, Eggplant - comfrey, chamomile, valerian (one part cottonseed meal, one part rock phosphate, one part bone meal). Chicken manure beneficial because of phosphate.

Potato - clover, comfrey, alfalfa, seaweed, buckwheat, dandelion (one part rock phosphate, two parts granite dust).

Pumpkin, Squash - comfrey, chamomile, valerian, buckwheat (one part rock phosphate, one part woodash, one part cottonseed meal).

Radish - shepherd's purse, dandelion (one part cottonseed meal).

Wheat, Oats, Barley - clover, alfalfa, dandelion, horsetail herb.

Many gardeners know that if we wish to eradicate a certain weed in our gardens we should make a diligent attempt at turning the weed into compost. When the soil receives the weed in the form of a compost it no longer feels the need to make the weed manifest physically. If thistles plague you, gather as many thistles as possible and ferment them into a tea. This tea sprayed on the soil will disincline it to produce thistles and give the soil the potassium it needs in a dynamic, life-filled form. Thus weeding and feeding involve the same activity and plant life flourishes. At the heart of this alchemy in nature is balance, the balance of the elements. To truly balance the forces of the elements in our gardens is an art like painting or poetry. Here the gardener and the artist can merge into the great rhythms of nature. The poems in the final section of this book are directed towards drawing the activity of the Muse into our work with the soil.

Earth, air, water and fire can then be metamorphosed into activities that seek to connect the artistic, cosmic drama of the seasons to our earthly workings in the garden.

Cosmic and Earthly Rhythms

Verses

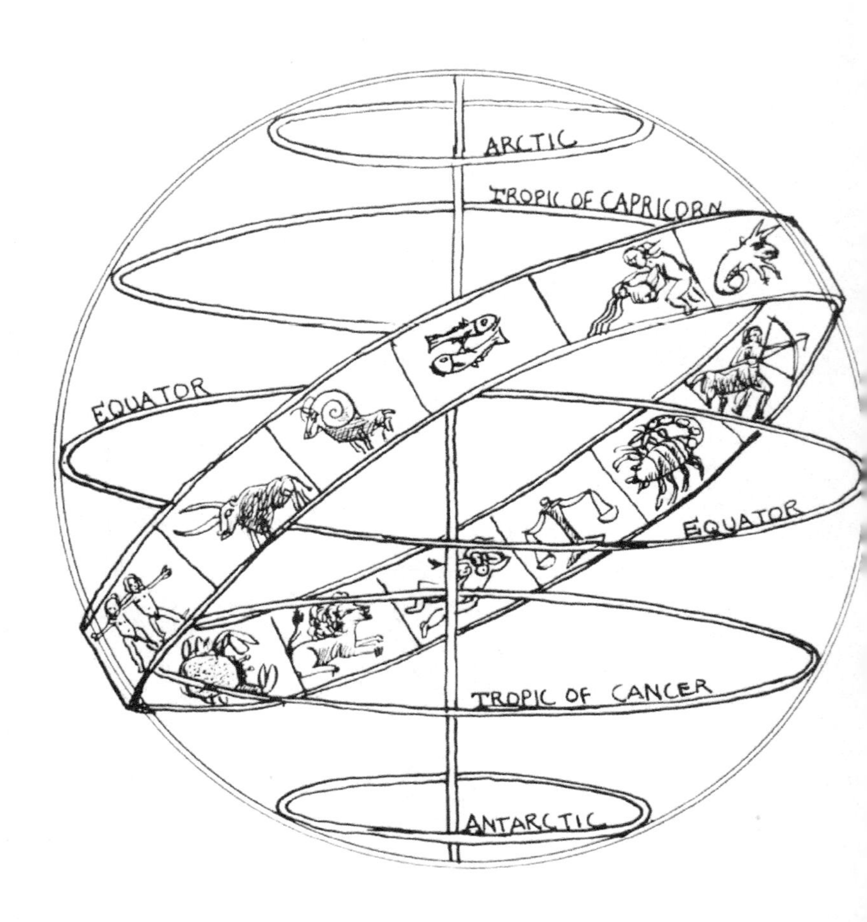
ARCTIC
TROPIC OF CAPRICORN
EQUATOR
EQUATOR
TROPIC OF CANCER
ANTARCTIC

Beyond the planets' circling run,
Around the life-bestowing Sun,
The void of space is filled with stars
That kindle Earth's life from afar.

And to the ancient men their forms
Appeared like animals with horns,
Or fish, or crabs, or girls with charms,
Lions, or men with bows at arms.

Each constellation gives the Sun the force
To move men and animals in evolutionary course.
And as the Sun moves within his harnessed track
He flies before this holy Zodiac.

The Sun's bright face serves as a lens
To focus starry light into the world of men.
This cosmic sunlight, quickening in Earth,
Gives men and animal forms their birth.

It's through the stars the souls of men appear,
And to the stars return when finished here.
Know the stars to know the animals and plants;
The shining heavens guide the living dance.

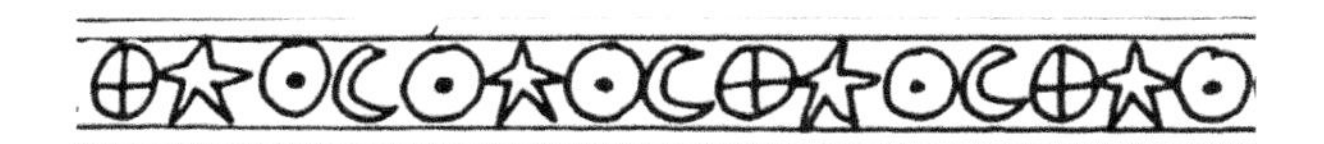

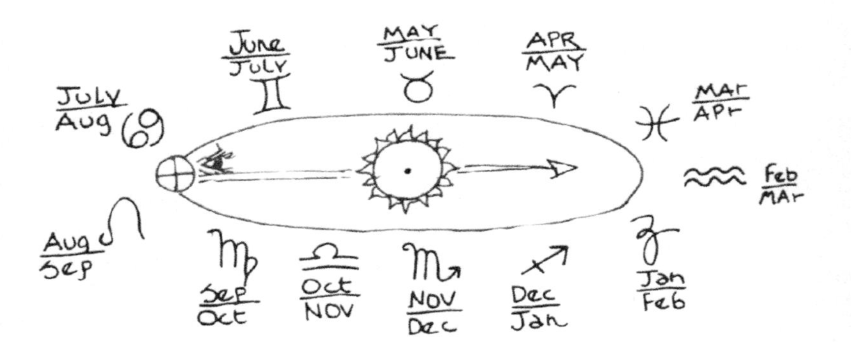
June
July
MAY
JUNE
APR
MAY
July
Aug
MAr
APr
Feb
MAr
Aug
Sep
Sep
Oct
Oct
NOV
NOV
Dec
Dec
Jan
Jan
Feb

To find the stars of the Zodiac,
Look to the south within the track
The Sun will travel in the day;
These stars are spread along his way.

The summer constellations lie
Above the trees, and two hands high;
The winter constellations found
Are four or five hands off the ground.
The summer nighttime stars become
The winter stars behind the Sun.
The winter nighttime stars appear
In daylight as the summer nears.

The constellations change in size
As they appear to earthly eyes.
In general they're two hands wide;
From east to west they number five
Or six depending on the season,
Look above your south horizon.

The stars by day are faint and dim
With the Sun in front of them.
Sunlight coming thus from space
At noontime marks his highest place,
While monthly, slowly passing by,
A constellation in the sky.

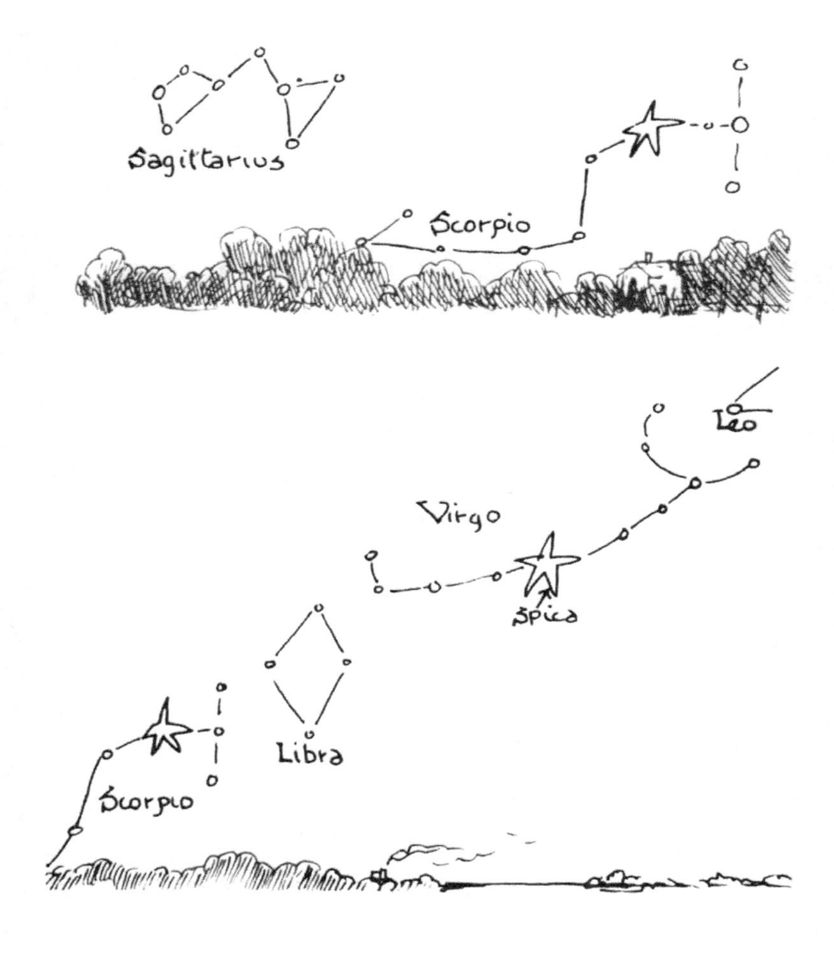
Sagittarius
Scorpio
Leo
Virgo
Spica
Libra
Scorpio

In summer when the nights are clear
Brave Sagittarius appears;
In the south a dipper and a bow are seen,
Hovering four fingers above the leafy green.

Westward the archer aims his bow
At the curving line of stars called Scorpio,
With glowing red Antares in his neck,
Moving westward, never looking back,

From Scorpio, but higher and to the west
Is the diamond shape of Libra, dim at best.
Sixteen fingers farther up the sky
Find Spica, brilliant star in the Virgin's thigh.
The Virgin stretches out in ease complete,
Her upraised arm just touching Leo's feet.

In fall the stars seem dim and hid by haze,
But Capricorn is there within our gaze.
To the east of vivid Sagittarius
A large inverted triangle is Capricornus.
East of the goat, we see four crossing stars:
Aquarius, pouring water from his jars.

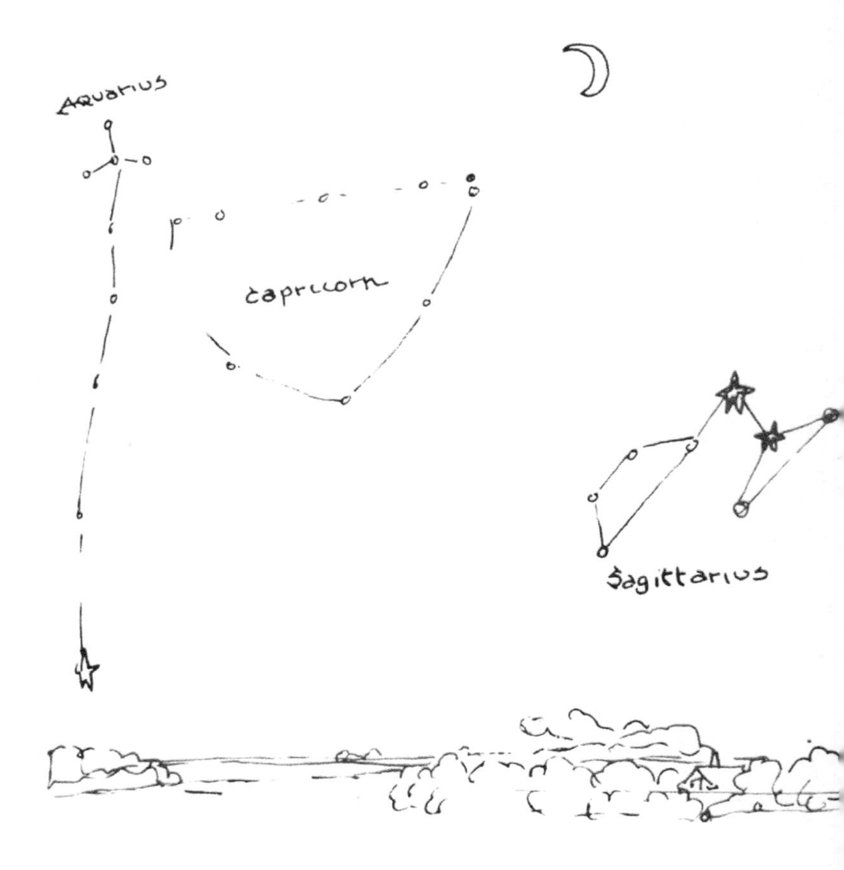

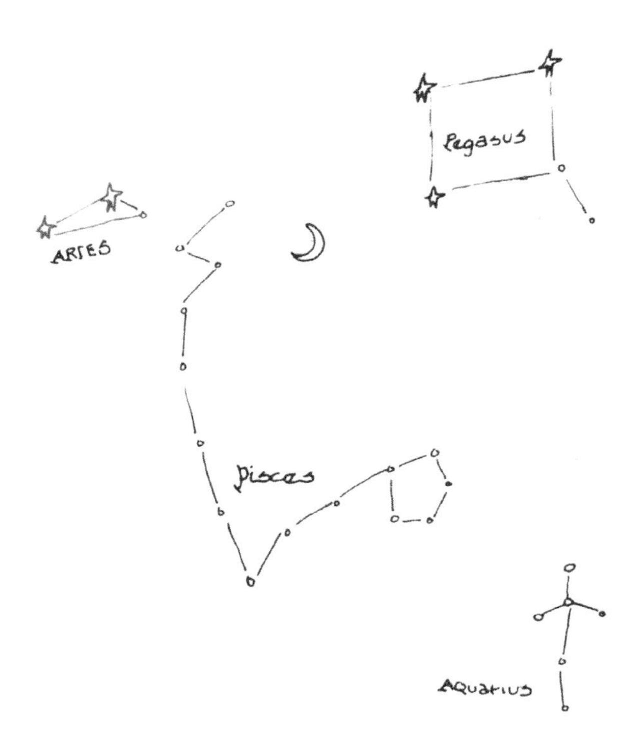

Above Aquarius, and overhead, we see
The large square Pegasus, circumpolar steed
Though not not himself within the Zodiac
He points to Aquarius, with Pisces at his back,
The western fish of Pisces, starry pentagram,
Is found ten fingers east of cross Aquarian.
The eastern fish of Pisces is astride Pegasus' tail
And further east is Aries, a triangle small of scale.

East of Aries, the Pleidaes, in shoulder of the bull:
These six stars in cluster signal coming of the chill.

With winter comes Orion, great hunter in the sky;
Three bright stars in a line point to Taurus' eye.
Aldebaran is the star on the east side of his face;
His horns point out to Gemini, the twins of stellar space,
Castor linked to Pollux, high up and overhead:
When bright and clear at eve, earth's cold and plants are dead.

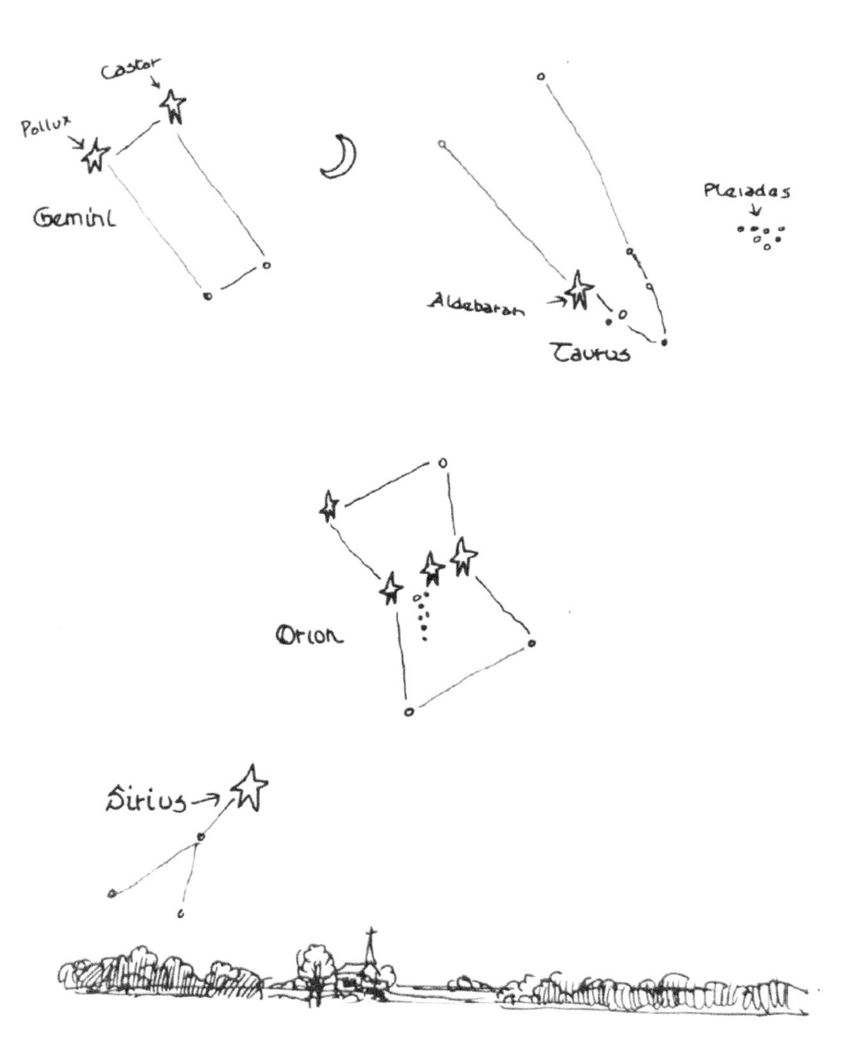
Castor
Pollux
Gemini
Pleiadas
Aldebaran
Taurus
Orion
Sirius

As Orion nears the setting Sun,
Winter cold is almost done.
The hook of Leo shines on high,
Far to the east of Gemini.
Cancer lies there in between,
A small, faint square just barely seen.

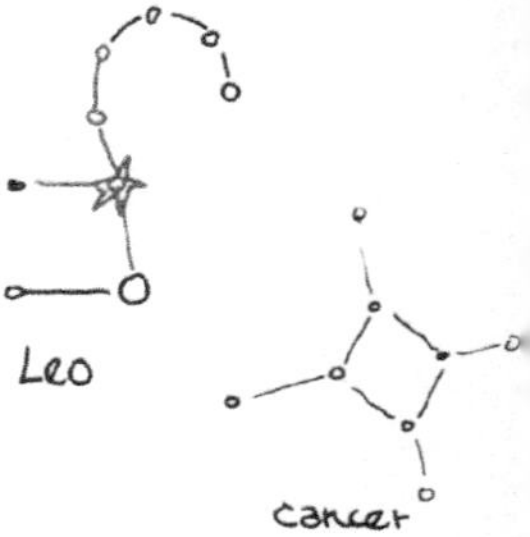

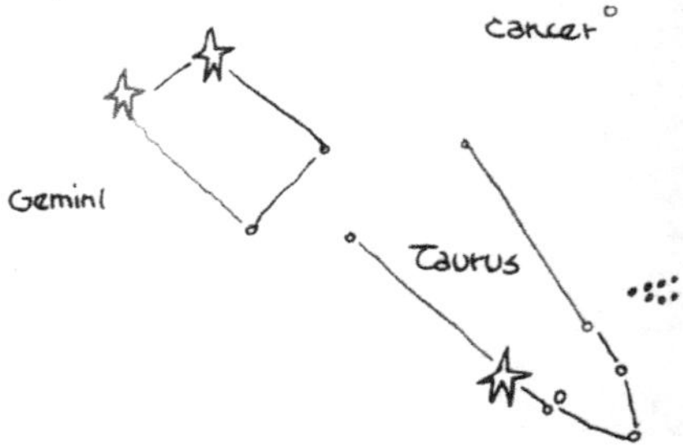

At Leo's tail the Virgin lies,
Sprawling through the springtime skies.
As Virgo passes to the west,
The summer heat and haze suggest
That Scorpio is in the sky,
And another year has passed us by.

Planets

Just like sister and like brother,
Planets are peers of one another.
Their influence as they move along
On Earth conditions is very strong.
The inner ones work in the atmosphere;
Air and water change and reappear.

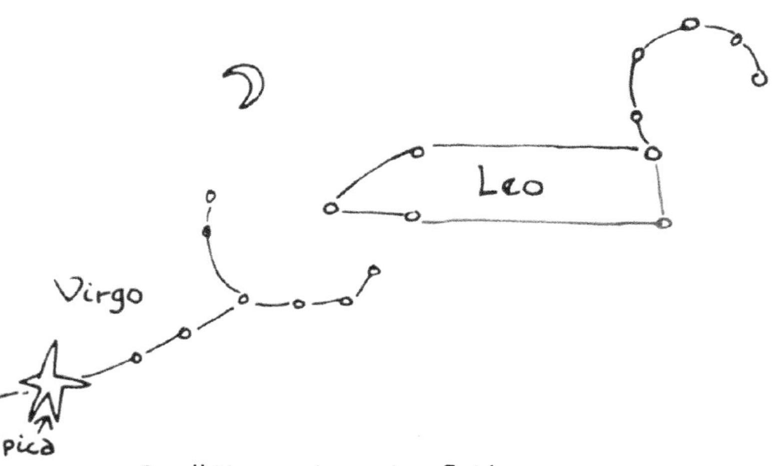

Small Mercury is moving, fluid grace;
He lends his hand to any change of place.
Venus, harmonizer of the air,
Clears the fronts and lets the Sun appear.
The sister of Earth, the Mirror Moon,
Controls the tides to give the Earth its boon
Of fertile energy to form the plants,
As Moon and Earth around the Sun both dance.

Mars, outside our orbit round the Sun,
Gives spatial forms to Earth through warmth and rain.
Jupiter, great planet of the moving air,
Gives winds and plastic force to earthly spheres.
Saturn aids the element of fire;
That the Spirit may evolve is His desire.

The planets wandering through solar space
Are sensitive to any change of place;
Among themselves great energy is passed,
Exchanges that are strong but do not last.

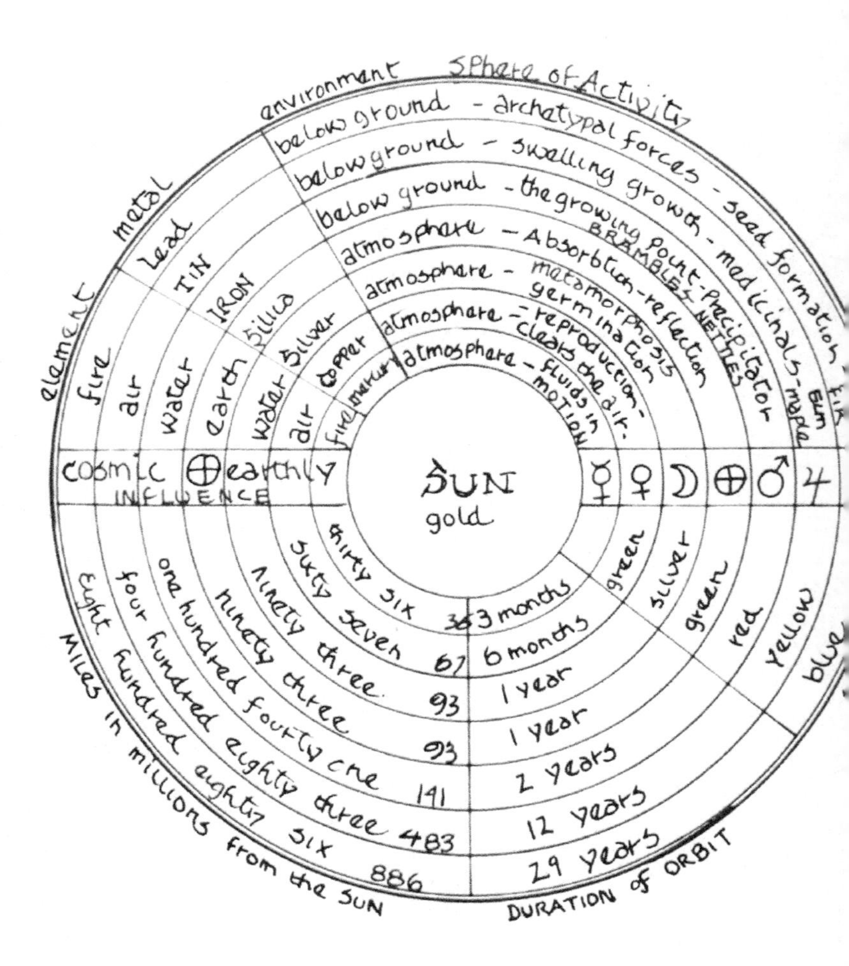

environment
Sphere of Activity
below ground - archetypal forces - seed formation
below ground - swelling growth - medicinals - maple
below ground - the growing point - precipitator
BRAMBLES NETTLES
atmosphere - Absorbtion - reflection
atmosphere - metamorphosis germination
atmosphere - reproduction - clears the air
atmosphere - fluids in MOTION
element
metal
fire
air
water
earth
water
air
fire
Lead
TIN
IRON
Silica
Silver
copper
cosmic
earthly
INFLUENCE
SUN
gold
green
silver
green
red
yellow
blue
thirty six
sixty seven
ninety three
ninety three
one hundred fourty one
four hundred eighty three
Eight hundred eighty six
Miles in millions from the SUN
36
67
93
93
141
483
886
3 months
6 months
1 year
1 year
2 years
12 years
29 years
DURATION of ORBIT

We gardeners on Earth are wise to keep
A watchful eye on planets, in the deep
Of space, that change the Earth's complexion
And alter it through interaction.

When planets seem to touch up in the blue,
And in a line are more than two,
End planets then will polarize,
While the middle one is vitalized.
They lie in opposition in the skies.

When Earth is in the middle of the row,
Tension from the ends will start to flow,
For planets in the Earth are subtly mixed
And so a mellow harmony is fixed.
Oppositions help the gardener work his tricks.

But when the Earth is found at either pole,
The Earth is irritated in Her soul.
Plant life suffers in the stress,
And working the Earth will bring regret.
Conjunctions make the Earth upset.

The Moon's in opposition at the full;
Conditions then are mellow, moist and full.
The Moon is in conjunction at the new;
Conditions then for growth are very few.

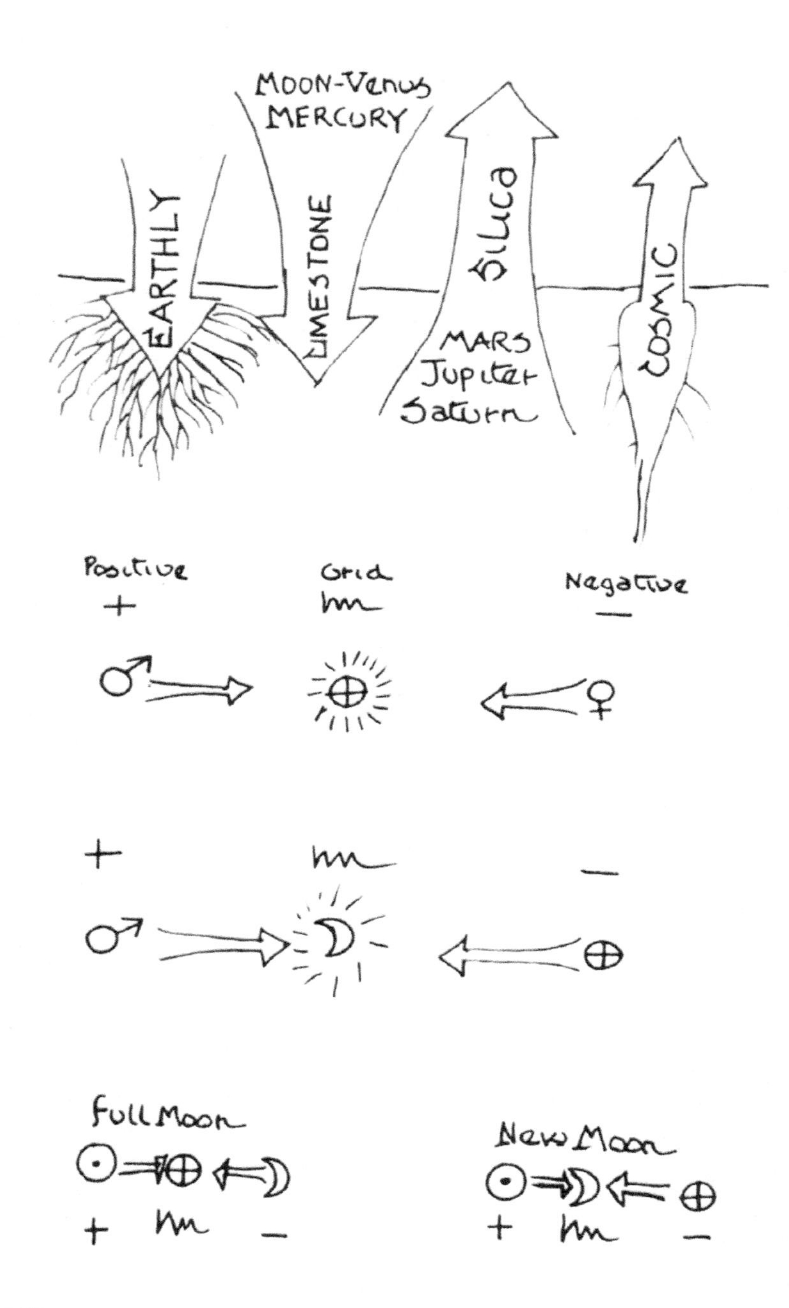
MOON-Venus
MERCURY
EARTHLY
LIMESTONE
Silica
MARS
Jupiter
Saturn
COSMIC
Positive
+
Grid
Negative
—
+
—
Full Moon
+
—
New Moon
+
—

For half the year the Sun's great light
Rises, building strength and might.
From winter solstice, darkest hour,
The Sun for six months grows in power.

At equinox in spring the Earth
Responds by giving plants their birth.
The green tips rise just like the Sun;
Earth's outbreath now has just begun.

Green with rising, vital power,
The leaves ascend to meet the flower,
Until amid the light-filled days
In Gemini the great Sun stays
His rising quest and calls an end
To running high; then he descends.

Once past the summer solstice peak,
His power lessens, week by week,
Fruits are set and seed is formed;
The plant world works as if alarmed.
The ripened fruits start to decay,
As solar power goes away.

At the equinox in Fall,
Earthly Virgo ripens all
The crops, before the winter chill
Brings the darkness deep and still.
And so through darkness to December,
The Sun is like a dying ember.
The Earth withdraws Her jewels green
To sleep within their roots and dream
Of cosmic joy and harmony when
The life-giving Sun shall rise again.

GEMINI
SUMMER SOLSTICE
Ripening
HOT
GROWING
WET
OUT BREATH
FLOWERING
AIR
PISCES
SPRING EQUINOX
GERMINATE
WATER FIRE
DECAY
FALL EQUINOX
VIRGO
EARTH
REST
POTENTIZING
COLD
INBREATH
DRY
ABSORPTION
WINTER SOLSTICE
SAGGITA

The Sun, while traveling before the stars,
Bestows the four elemental powers.
To change the seasons on the Earth,
And through the elements give birth
To processes of growth and change
In plants and animals, the range
Of earth and water, fire and air
Are woven in the earthly sphere.

O song of earth and water wed,
The earthly pole of muck and mud!
Of worms and roots these elements sing;
Fertility is what they bring.
The earthly, mineral-forming force
Of crystals needs the water course,
Chemically combined to make
The plants from minerals and take
Them back again when they expire,
Or change them into something higher.

The air and light then change the shoot
Of water coming from the root.
They photosynthetically produce
The leaf, a membrane filled with juice.
This airy structure, broad and thin,
Then pulls the sunlight deep within,
To penetrate the earthly gloom
Of water mixed with dirt and loam.

The air refines the earthy mould
Into a structure that can hold
The insect in an act of love.
So flowers nod far up above
The Earth, and wait for wind to move
Or wings to stir them into love.
The insect comes to bring the fire
That raises plant life one step higher.
The plant supports the honeybee
And her evolved society.
The fire of thought the insects give,
So plant and insect both can live.

Within the flower life is changed;
The flower dies, the seed remains.
The seed's the ash of life, condensed
To foster generations hence,
Produced of fire contracted dry,
Removed from life, dead to the eye.
But placed into the living womb
Of earth, the seed can then resume
The dance that earth and water form
As life from death again is born.

Bee
FLOWERS
SILICA
AIR
BIRDS
SEEDS
PHOSPHOR
FIRE
EARTH
LIME
ROOTS
WORMS
WATER
POTASH
LEAVES
COW

In olden books of garden lore,
Moon phases are but one of four
Great rhythms moving watery sap;
In plants they rule the growing tip.

The phases move the watery tides
Through the seas and land besides.
At full Moon, Earth is fertile; then
At dark Moon she is weak again.
The synodic Moon that waxes, wanes
Alters horizontal planes.

By bringing moisture at the full,
The Moon makes active earthly soil
But as she moves into the dark,
The drying Moon retards the spark
Of energy within the breast
Of Mother Earth, and so plants rest.
At the quarters there is strain
Within the atmospheric plane;
Weather fronts develop fast,
Storms are strong but do not last.

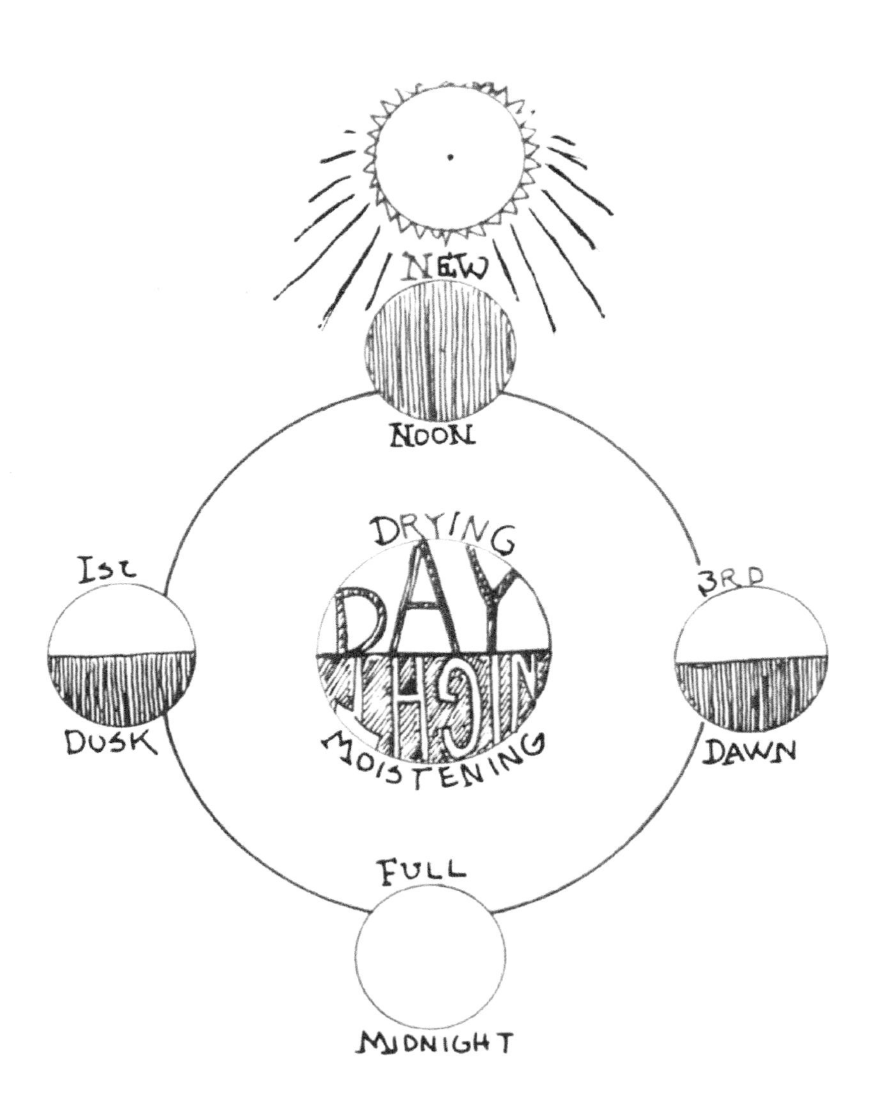
NEW
NOON
DRYING
DAY
NIGHT
MOISTENING
1st
DUSK
3RD
DAWN
FULL
MIDNIGHT

The Moon runs high and she runs low,
As well as growing dark and full.
Vertical forces from below
Rise and fall with the lunar pull
For two weeks the Moon descends;
This keeps her sap in the root ends.
Until her arc is above the trees,
Growth is slack, plants are at ease.
For two weeks then the Moon will rise;
This pulls the sap up toward the skies.
Until her arc is overhead,
Plant life grows and flowers spread.
The full Moon waxes and it wanes.
The Moon will rise and fall again
This makes the Earth grow strong and weak,
From rooty low to flowery peak.
If full Moon comes when she runs high,
The tips of plants grow rich and bright.
If full Moon comes when she runs low,
Fruits get ripe and Earth is mellow.
If dark Moon comes while she runs high,
Crops and herbs are sure to dry.
If dark Moon comes while moon runs low,
Earth is dry and growth is slow.

As the Moon does fall and rise, She passes
through the starry skies.
She is a lens for stellar power,
To form the root, the leaf, the flower.

CANCER
LEO
VIRGO
LIBRA
SCORPIO
SAGITTARIUS
CAPRICORN
AQUAR
TAURUS
Gemini
FRUITING - seeding - transplanting - harvest & use - dormancy - cultivate well - germinate - Rapid growth - FLOWERING
RIPENING - DISSOLVING - DECAYING - ROOTS GROW
VITALIZING - ENHANCING - REFINING - COSMIC
HORIZONTAL PLANE - EARTH GROWS STRONG
VERTICAL PLANE - PLANT TIPS GROW STRONG
MOON RUNS HIGH
MOON RUNS LOW

When Earth is turned by plow or hoe,
Star force is fixed to what you sow.
What's in the cosmos as you toil
Gets fixed into the earthly soil.
The sign the Moon does occupy
Will guide the seedlings' destiny.

Seed when sown in Moon of Earth
Grows strong roots of ponderous girth.
Seed sown in lunar water signs
Grows sappy leaves and running vines.
Seed sown in the Moons of air
Grows flowers colored, fragrant, rare.
Seed sown in the Leo sign
Grows grains and pulses, full and fine.

The final rhythm of the four
Finds the Moon both near and far.
At apogee she's far away;
If seed is sown upon that day
It tends to rot down in the row;
The germination is too slow.
At perigee the Moon is near,
And plant life grows as if in fear;
Life is short, plants bolt to seed,
If they are sown at perigee.

The rhythms of the Moon are four:
Close at hand and very far;
Running low and running high;

Dark or full in nighttime sky;
And running through the starry track,
The Moon reflects the Zodiac.

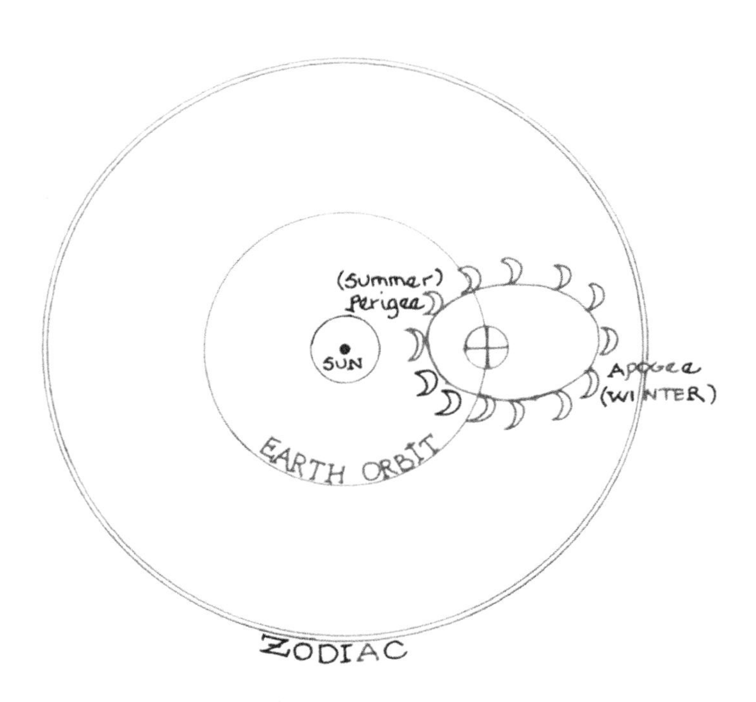

The Gardener's Year in Verse

Pisces

In Pisces, Earth starts breathing out;
When seeds are sown, they quickly sprout.
Pisces brings the cold spring rain
That helps bring seeds to life again.
Sow in compost, on days for leaves,
Leeks and onions, sets or seeds.
Sow carrots, turnips, parsnips, beets;
To germinate them, soak the seeds.
Sow cabbage, tomatoes and peppers within,
And plant them out when Taurus comes.
Peas, when sown in the Leo Moon,
Fill the pods as hard as stone.
When the earthworm moves about,
Plant the hardwood cuttings out.
To keep your scion wood asleep,
Dig a trench that's two feet deep;
In sawdust and dampened soil enfold,

To keep them moist but airy and cold
They should sleep until it's warm
Or else the scions come to harm.

Rootstocks grown last year are strong.
Cleft graft the cherries before too long.
Set your grafts as Moon ascends;
Root or leaf days help them mend.
When setting scions into root,
With hopes to build the fruiting shoot,
You give them four buds each, no more;
Let one the leader be for sure.

Whip and tongue the apple and pear;
Tight cambium contact aids repair.
Wrap them, wax them, join them tight.
In Taurus, buds will see the light.

Take a bladder from a stag;
With yarrow flowers stuff the bag.
Hang it in full summer Sun,
And take it down when summer's done.
Bury it through winter cold,
Then take it out, when one year old,
And add to compost in the heap.
The sulfur forces then will keep
The compost full of cosmic power
That quickens Earth to fruit and flower.

Until your stocks show signs of life
Keep well hid the grafting knife.

Aries

Thin your seedlings in each row;
Take the weak, the strong will grow.
Hoe the roots in earthy signs,
Hoe in fire for fruits and vines,
Hoe for flower in signs for air,
Hoe for seed in Leo fair,
Hoe for leaf in watery moon;
Hoeing can't be done too soon.
Plant potatoes, one eye alone,
Near apogee, in sandy loam.
On seed days plant, if eyes you need.
On root or leaf days, plant for food.

Prune the fruits as Moon descends;
Take water shoots and brittle ends.
Best pruning is when leaf has sprung

But shooting branch has not begun;
Seven spurs per branch should stand;
Prune the least amount you can.
Spray watered skim milk on fruit trees,
As blossoms open for the bees.
To help fruit grow, spray nettle teas,
With seaweed or comfrey, on the leaves.
Spread some compost on the root
To strengthen next year's fruiting shoot.

With silica fill a cow horn sound,
And bury it within the ground.
The summer earth will change the quartz
And concentrate the cosmic source
Of lightfilled metamorphic force
Into the contents of the horns.
If placed in water and then stirred
(An hour is the time preferred),
Then sprayed upon the seedling plants,
This mixture makes the leaves advance.

Taurus

Sow your beans, soy, lima, pole,
And point their eyes down in the hole.
Sow the fruit crops for the fall,
Wax beans, squash, and melons all.
Transplant cabbage, broccoli, leeks,
As Moon descends, on days for leaves.
Transplant tomatoes and the peppers,
Sow the tender annual flowers.

Cut for softwood when it snaps,
As Moon runs high and into wax.
Gather cuttings in the morn;
Ascending Moon tips will be strong.
Take side branches, not the ends,
And plant them as the Moon descends.

June budding can at last commence,
As bark slips for three months hence.
Use upright limbs to be your stock,
So water will not rot the bark,
Use dormant buds from scion wood;
Cuttings from last fall are good.
If leaves fall off, you need not fear;
The bud is rooting for next year.

Take the intestines from a cow;
Pick the chamomile flowering now.
Make "sausages" filled with the bloom;
Bury them shallow in fertile loam
Through the winter's icy glow
So they can freeze near sunlit snow.
Sprinkle them in springtime's heap;
Through calcium, they work to keep
Potassium and phosphorus
In balance, so fruiting's prosperous.

Gemini

On Gemini's bright and airy days,
Pick herbs and they will dry with ease:
Ascending Moon, they're vitalized;
Moon near the dark will help them dry.
Picked for leaf, herbs loose their power;
Use fire or air signs before they flower.
The scent stays in before the dawn;
Dry herbs in shade, avoid the Sun;
When crisp but green the herbs are done.

Firewood cut during Gemini's reign
Dries out quickly to your gain.
If taken green, make the bark undone,
Then lay logs out in the summer Sun.
Wood for building seasons best
Under water two weeks first.

If onions sown from sets now fail,
Pick in dry Moons; they'll store well.
Left in the ground through Cancer rains,
Stored onions rot before spring comes.
If sown from seed let onions go;
Harvest them in dry Leo.
Next spring green onions are a cinch,
From onions smaller than an inch
Leeks are best in Fall and Winter,
Planted when thick as little finger.
Virgo transplants are the best;
Move the biggest, thin the rest.

Give the carrots some wood ash,
And borer flies will go right past.
Fall crops sown as Gemini wanes
Transplant well in Cancer rains.
For the peach and nectarine,
Early budding can begin.
Use dormant buds from this year past,
As moon runs high and into wax.

Cancer

Watery Cancer makes plants grow.
Move kale and cabbage into rows.
To fill the spaces in between,
Sow some lettuce and late beans.
Sow late radishes and beets.
Transplant broccoli and leeks.
Sow the legumes, to be sure:
Alfalfa, clovers, green manure.
The Moon in Taurus makes roots strong
To help a pasture come along.
If seed is sown to grow fine hay,
Sow it on a flower day.

As Cancer sunlight fills the skies,
Pick black and blue and dewberries.

Morning is the time to pick;
As noon descends, they ripen quick.

When the bark slips on the wood,
Use buds asleep at the base of a shoot
And budsticks can be set with ease.
As Moon descends, you gather these.
The rising Moon helps buds to lock
When they are set in rooted stock,
As leaves pull off, the bud is ripe;
Next year a branch will seek the light.
Cut the stock rind, not the wood;
The north side of the stock is good.
Place the buds in "T"-shaped cuts,
Wrap them, wax them, that's enough.

Pack a cow horn with cow dung.
Bury this away from Sun
In fertile soil through Winter cold;
The horn will cause the dung to hold
The earthy forces in its mass,
Forming humus that will last.
Stirred in water for an hour,
This preparation helps Earth's power
When sprayed in fall or with the sower.

Leo

In Leo heat in morning hours
Gather seed for herbs and flowers.
Pick seed in signs of the crop you desire,
Root seed in Earth, fruit seed in Fire,
Leaves in Water, flowers in Air;
The seed will follow suit next year.
While Moon runs high, seed strength remains.
While Moon runs low, the seed power wanes.

With Leo in the daytime sky,
Cooking herbs are sure to dry
As fiery or airy Moons go by.
Rain pails filled as full Moon nears
Will help dry seedlings, never fear.
Divide and plant the strawberries,
So they can set without a freeze.

Pick grapes and peaches, elder, pear;
Leo's heat makes nectar rare.
Fruit that's picked as Moon ascends
Keeps until the season ends.
Descending Moon will ripen sweet
Fruit for canning or to eat
Prune old fruits as leaves first drop;
Next year they might give a crop.

With trees and evergreens to deal,
Take half-ripe cuttings with a heel
Plant only side-shoots in coarse sand;
In cold frames one whole year they'll stand
Keep them moist, protect from cold,
Avoid strong Sun 'till one year old.

Cut some stinging nettles down
And pile them up into a mound;
(Careful that you don't get stung!)
Cover up the mound with dung.
Wait a year and they will mold
Into a humus that's like gold.
Added to your compost pile,
This prep will make all plant life smile.

Leo transplants take the heat,
Wait till Virgo moist and sweet.

Virgo

Descending Moon helps mend the roots,
The waning moon checks water shoots.
Avoid a moon that's high and waxing;
The quickened growth will be too taxing.
When planting fruit trees always put
Some bones and nails beneath the root.
If your land is on a hill,
To the south your garden till.
But plant your orchard to the north
So early thaws don't call buds forth.
If grapes on mulberries are put,
They both will bear a better fruit.

Hardwood cuttings with a heel,
Placed firm in compost, next year yield.

Plum and quince will soon take hold
If kept from sun and wind and cold.

Apples, pears and cherries die;
Save them till next spring comes by.

Find a skull that's sound and whole;
Put crumbled oak bark in the hole
And fill it up; then plant it where
Water will soak it for a year.
Add the contents to your heap.
This preparation works to keep
All plant life healthy, firm and strong,
For calcium is helped along.

Divide in Virgo all the roots,
Asparagus, iris, bamboo shoots,
Rhubarb, comfrey, clumps of herbs;
The rooting forces Virgo serves.
To plant spring bulbs for flowers fair,
Work on days that rule the air.
Add some bone dust and wood ash
For flowers that are unsurpassed.

If long-lived tree or bush you move,
Use Virgo Sun, with a water/earth Moon
When moving trees from place to place,
Mark well which side to north they face,
And plant them in the rows this way
To save them stress etherically.

Libra

One to rot and one to grow,
One for the sparrow, one for the crow.
Sow winter grains for body's needs;
When summer-sown, you grow good seed.
Shallot, garlic, onions for fall:
Eat the large cloves, plant the small.
Onion seeds may now be sown,
But in cold ground, when warmth has flown.
Plant them in a compost trench;
When they're frozen, you can mulch.
As frost knocks all the carrots down,
Take the strongest from the ground;
Store in damp sand, do not freeze;
Replant next spring to get good seeds.

Potatoes store, moist, dark and cool,
With carrots, cabbage, and beets as a rule.
But onions, yams and squash have needs
For cool, dry storage, like all seeds.
Eat leeks and carrots in the fall;
Save squash and onions for the thaw.
Late in winter, the soup pot
Needs vegetables still free from rot.
To keep the winter crops till thaw,
Mulch them only with dry straw.

When moon runs low, put compost down
To build the humus in the ground
Apply #500 in the fall,
Cover with leaves that start to fall.
Scatter wood ash, clay, and lime;
Crushed rock and burned bone are fine.
Feed bushes, trees and bramble fruits
To help the next year's fruiting shoots.
When soil has frozen, but not too deep,
Mulch carrots, parsnips and kale to keep.
In sandy soil the roots stay sound,
But they will rot in clayey ground.
Keep some lettuce under glass,
Dry, cool and airy while frozen fast.

Scorpio

For fruit tree health without complaint,
Cover trees with B. D. paint.
To guard the bark from frost and sun,
Make sure each branch is wholly done.
Take one part blood that's dry, and then
Add two parts Fuller's Earth; again,
Add three parts compost, very fine;
Mix these with water, stir; assign
Four parts fine clay—helps paint to stick!
For young trees, make it very thick,
Paint the trees from trunk to twig,
So they will flourish and grow big.
If paint corrodes before winter's end,
Add two parts dry milk to your blend.

One-half part wood ash adds some strength
And keeps the parasites at length.

Take dormant hardwood cuttings now;
A callus on the cut should grow.
Use side branches, keep the heel;
If planted now they rot and peel
Most cuttings need the light to grow,
So keep them bundled through the snow.

Store them heeled into the earth;
Don't let them freeze or die of thirst.
In Pisces, water gives roots birth.

Break up clays so frost can work
To turn each clod into fine dirt.

Bury next year's garden seeds
In fertile soil, away from weeds.
Put them into airtight jars;
The Earth will fill them with the stars.
Roots, and fruits and leaves will thrive
But peas and beans cannot survive
Throughout the winter, without air;
To keep them free from rot, take care.

Sagittarius

In the Archer's sign, fruit scions cut;
Take branches short in node, and stout.
Use this year's shoot with fattened bark
While Moon runs high and into dark,
Cut on days for root or leaf;
Store in damp sand two feet deep.
Before you cut for scions, please
Wait two days without a freeze.
Store scions cold so they will keep
Until the rootstocks wake from sleep.
Buds on scions should number four;
Pinch them out to one next year.
For top grafts helping old trees bear,
Leave seven buds reduced to four.
When the rootstocks show a leaf,
The scions all should be asleep.

If root we need, then root we sow;
To form the root, in root we hoe.
Pick roots on root days, when moon's low.

Before the corn is sown for seed,
Make sure the patch in Leo you weed;
But when the corn is getting high,
In Capricorn the hoe bring by.
We sow for leaf the cabbage heads
And cultivate in leaf besides,
But harvest on a flower day
So cabbages stay firm and dry.
The fruits need heat to make them sweet,
And fire is needed to form seed.
If vines and fruits you so desire,
Sow and cultivate for fire.
But sown with Leo in the sky,
Fleshy beans would be too dry.
For fat snap beans without big seed,
The Ram and Archer fill the need.

Capricorn

When the Moon is new to full,
Timber fibers warp and pull.
Wood cut while the Moon is dark
Builds houses tight as Noah's Ark,
When the Moon is new to full,
All the plants can drink their fill;
When the Moon is full to dark,
Plants dry out and form their bark.
When the Moon is new to fat,
Trees when moved will grow too fast;
When the Moon is full to new,
Trees will last, but they'll grow slow.
When the Moon is new to round,
Vegetables grow quick and sound.
When the Moon is full to new,
Roots get tough and fruits are slow.

The Quarter Moon, the Quarter Moon,
Brings the rain and wind and gloom.
Summer's rain stays in the cup
If February's new Moon's tips are up.
If the horns point to the side,
Rains come late in summertide.

When Moon runs high, it's harvest time
For crops that grow up on the stem.
When Moon runs low, then gather all
The roots to transplant; they'll do well.
When Moon runs high, cut brush, take slips;
When Moon runs low, then prune the tips.
When Moon runs high, set grafts, take seeds.
When Moon runs low, take fruit, kill weeds.

Aquarius

When the new Moon's tips are soft,
Some wet weather's not far off.
When tips are sharpened to a point,
Strong winds will howl all through the night.
When darkened Moon is clearly seen,
No rains will come to intervene.
Aches and pains and chairs that squeak
Foretell the rain, of rain they speak.
If in the fields the sea birds stand,
Storms come from sea and towards the land.
But when the geese move out to sea,
Good weather there will surely be.
When the fowl scratch together,
Farmer, you may have foul weather.
And when all birds and chickens roost,
Farmer, then put on your boots.

Soft moon rain, and red moon blow.
Bright Moon, fair a day or so.

When in spring the ant house builds,
Then goes the farmer to his fields.
When all insects sting and bite,
Then white leaves show a storm in sight.
When the insects start their swarming,
Soon the weather will be warming.
And when the bees stay home all day
A storm is sure to come your way.
When spider spins her silken thread,
Expect fair skies in days ahead.
Before a storm, the ants observe;
They walk in lines without a curve.

Weather Vanes

If the "wooly bears" are red,
A mild winter is ahead,
But if the "wooly bears" are black,
Winter will be cold and bleak.

With ringed Moon and wind that shifts,
The snow will soon lie in deep drifts.

Thunder in the morning,
All day storming.
Thunder at night,
Travelers delight.

Morning dew means weather fair;
When dew departs, a change beware.
Wind before rain,
Fair weather again.

When clouds appear scratched by a hen
Work to get your hay put in!

Mackerel sky, mackerel sky,
Dry turns wet, wet turns dry!

When clouds look just like wedges
Cold winds shake the trees and hedges.

When wind lies to the South or East
Tis not fit out, for man nor beast.
But if toward the North or West it lay,
Cold and bright it is today.
And when a winter snow remains,
A Southwest wind will bring warmer air
For Southern wind tells of a change.

References

Findhorn Community, The. *The Findhorn Garden*.
New York: Harper and Row, 1975.

Gregg, Richard B. *Primer of Companion Planting*.
Wyoming, RI.: Bio-Dynamic Literature, 1981.

Grohmann, Gerbert. *The Plant*.
London: Rudolf Steiner Press, 1974.

Lievegoed, B.C.J. *The Working of the Planets and The Life Processes in Man and Earth*.
Clent, England: The Experimental Circle of Anthroposophical
Farmers and Gardeners, 1972.

Pfeiffer, E.E. *Weeds and What They Tell*.
Wyoming, R.I.: Bio-Dynamic Literature, 1981.

Philbrick, Helen and Richard B. Gregg. *Companion Plants and How to Use Them*.
Old Greenwich, Conn.: The Devin-Adair Company, 1966.

Rodale, J.I., ed *Encyclopedia of Organic Gardening*.
Emmaus, Pa.: Rodale Books, 1975.

Schwartz, Sherry, ed. *The Kimberton Hills Agricultural Calendar*.
Camphill Village, U.S.A, Inc.

Schwenk, Theodor. *Sensitive Chaos*.
New York: Schocken Books, 1975.

Steiner, Rudolf. *Agriculture*.
London: Bio-Dynamic Agricultural Association, 1977.

Storl, Wolf. *Culture and Horticulture*.
Wyoming, RI.: BioDynamic Literature, 1978.

Thun, Maria. *Working on the Land and the Constellations*.
Sussex, England: Lanthorn Press, 1978.

Thun, Maria and Matthias K. *Working With the Stars: A BioDynamic Sowing and Planting Calendar*.
Sussex, England: Lanthorn Press, 1981.

For the bio-dynamic sprays and compost preparations:

Josephine Porter
RD 1 Stroudsburg,
Pennsylvania 18360

About the Author

Dennis Klocek was born in 1945, in Shenandoah, Pennsylvania, and spent his youth in Philadelphia and in the mountainous coal-mining regions of his native state. His interests, from 1903-67 when he attended Glassboro State College, centered around art, art education, and the biological and earth sciences. After two years in the Navy, he taught art in the public schools of Mt. Holly, New Jersey, and did graduate work at Temple University's Tyler School of Art where he was a teaching fellow from 1970-72. For the next seven years he taught drawing and painting at Brookdale Community College in Lincroft, New Jersey, and it was during this period that he did the Illustrations for the edition of Goethe's ***Metamorphosis of Plants*** published by Bio-Dynamic Literature in 1978. After this, he says, "my teaching took a decided Goethean turn," leading to his giving a course in Goethean morphology at Brookdale. He began reading such authors as Hermann Poppelbaum, Theodor Schwenk, Walther Cloos (***The Living Earth***), Maria Thun, and E. E. Pfeiffer, as well as issues of ***Bio-Dynamics***, while at the same time studying Rudolf Steiner's agriculture lectures and pursuing bio-dynamic gardening with great interest. "Gardening then became a focus for meditations on the elements, and the idea for the Book of Moons developed." A reading of Goethe's ***Faust***, and of Wolf Storl's ***Culture and Horticulture*** further crystallized, in different ways, the imaginative concepts behind ***Moons***. With this impulse alive within them, he and his wife and three sons left the east coast for Eugene, Oregon, where he taught Biodynamics in the Cascade Valley Waldorf School and a course in "Gardening for the Almanac" to the adult community there. He is now taking the teachers' training course at Rudolf Steiner College in Fair Oaks, California, where he will also be teaching some Bio-dynamics. He plans eventually to teach in a Waldorf School where he can utilize both his scientific and artistic training and continue to communicate to others his love for the world of plants.

Illustrations and decorative art by Dennis Klocek

For Farmers, Healers, Educators

Also by Dennis Klocek

Sacred Agriculture
The Harmonies of Storms
Climate: Soul of the Earth
The Seers Handboook
Weather and Cosmos
Drawing From the Book of Nature
and more.

See dennisklocek.com for

Audio Lectures
Video Courses
Articles
Other Publications

Printed in Great Britain
by Amazon

57661631R00086